FOR REVIEW

S. Ejaz Ahmed
Book Review Editor
Technometrics

# New Theory of Discriminant Analysis
After R. Fisher

Shuichi Shinmura

# New Theory of Discriminant Analysis After R. Fisher

Advanced Research by the Feature-Selection Method for Microarray Data

Shuichi Shinmura
Faculty of Economics
Seikei University
Musashinoshi, Tokyo
Japan

ISBN 978-981-10-2163-3     ISBN 978-981-10-2164-0  (eBook)
DOI 10.1007/978-981-10-2164-0

Library of Congress Control Number: 2016947390

© Springer Science+Business Media Singapore 2016

This work is subject to copyright. All rights are reserved by the Publisher, whether the whole or part of the material is concerned, specifically the rights of translation, reprinting, reuse of illustrations, recitation, broadcasting, reproduction on microfilms or in any other physical way, and transmission or information storage and retrieval, electronic adaptation, computer software, or by similar or dissimilar methodology now known or hereafter developed.

The use of general descriptive names, registered names, trademarks, service marks, etc. in this publication does not imply, even in the absence of a specific statement, that such names are exempt from the relevant protective laws and regulations and therefore free for general use.

The publisher, the authors and the editors are safe to assume that the advice and information in this book are believed to be true and accurate at the date of publication. Neither the publisher nor the authors or the editors give a warranty, express or implied, with respect to the material contained herein or for any errors or omissions that may have been made.

Printed on acid-free paper

This Springer imprint is published by Springer Nature
The registered company is Springer Science+Business Media Singapore Pte Ltd.

# Preface

This book introduces the new theory of discriminant analysis based on mathematical programming (MP)-based optimal linear discriminant functions (OLDFs) (hereafter, "the Theory") after R. Fisher. There are five serious problems of discriminant analysis in Sect. 1.1.2. I develop five OLDFs in Sect. 1.3. An OLDF based on a minimum number of misclassification (minimum NM, MNM) criterion using integer programing (IP-OLDF) reveals four relevant facts in Sect. 1.3.3. IP-OLDF tells us the relation between NM and LDF clearly in addition to a monotonic decrease of MNM. IP-OLDF and an OLDF using linear programing (LP-OLDF) are compared with Fisher's LDF and a quadratic discriminant function (QDF) using Iris data in Chap. 2 and cephalo-pelvic disproportion (CPD) data in Chap. 3. However, because IP-OLDF may not find a true MNM if data do not satisfy the general position revealed by student data in Chap. 4 (Problem 1), I develop Revised IP-OLDF, Revised LP-OLDF, and Revised IPLP-OLDF that is a mixture model of Revised LP-OLDF and Revised IP-OLDF. Only Revised IP-OLDF can find true MNM corresponding to an interior point of optimal convex polyhedron (optimal CP, OCP) defined on the discriminant coefficient space in Sect. 1.3. Because all LDFs except for Revised IP-OLDF cannot discriminate cases on the discriminant hyperplane exactly (Problem 1), NMs of these LDFs may not be correct. IP-OLDF finds Swiss banknote data in Chap. 6 having six variables is linearly separable data (LSD) and two variables such as (X4, X6) is minimum linearly separable model by examination of all 63 models made by six independent variables. Revised IP-OLDF confirms this result, later. By monotonic decrease of MNM, 16 models including (X4, X6) are linearly separable models. This fact is very important for us to understand the gene analysis. Only Revised IP-OLDF and a hard-margin support vector machine (H-SVM) can discriminate LSD theoretically (Problem 2). Problem 3 is the defect of generalized inverse of variance-covariance matrices that causes a trouble for QDF and a regularized discriminant analysis (RDA). I solve Problem 3 that is explained by the pass/fail determinations using 18 examination scores in Chap. 5. Although these data are LSD, error rates of Fisher's LDF and QDF are very high because these datasets do not satisfy Fisher's assumption. These facts tell us serious problem that we had better re-evaluated the discriminant results of Fisher's LDF and QDF. In particular, we shall re-evaluate the medical diagnosis, and various ratings because these data have the same type of test data having many cases on the discriminant hyperplane. Because Fisher never formulated the equation of standard error (SE) of error rates and discriminant coefficient (Problem 4), I develop a 100-fold cross-validation for small sample method (hereafter, "the Method 1"). The Method 1 offers the 95 % confidence interval (CI) of discriminant coefficient and error rate. Moreover, I develop a powerful model selection procedure such as the best model with minimum mean of error rate in the validation samples (M2). Best models of Revised IP-OLDF are better than other seven LDFs using six datasets including Japanese-automobile data in addition to above five datasets. Therefore, we misunderstand I establish the Theory in 2015. However, when Revised IP-OLDF discriminates six microarray datasets (the datasets) in November 2015, Revised IP-OLDF can naturally select features. Although Revised IP-OLDF can make feature-selection naturally for Swiss banknote data and Japanese-automobile data in Chap. 7, I do not think it is a very important fact because the best model offers the useful model selection procedure for common data. Over than ten years, many researchers are struggling in the analysis of gene datasets because there are huge numbers of genes and it is difficult for us to analyze by common statistical methods (Problem 5). I develop a Matroska feature-selection method (hereafter, "the Method 2") and LINGO program. The Method 2 reveals the dataset consists several disjoint small linearly separable subspaces (small Matroska, SMs) and other high-dimensional subspace that is not linearly separable. Therefore, we can analyze each SM by ordinary statistical methods. We find Problem 5 in November 2015 and solve it in December 2015.

The book represents my life's work/research, to which I have dedicated over 44 years of my life. After graduating from Kyoto University in 1971, I was employed by SCSK Corp. in Japan as a system integrator. Naoji Tuda, the grandson of the second-generation general director Teigo Iba of Sumitomo Zaibatsu, was my boss and he believed that medical engineering (ME) is an important target for the information-processing industries. Through his decision, I became a member of the project for the automatic diagnostic system of electrocardiogram (ECG) data with the Osaka Center for Cancer and Cardiovascular Diseases and NEC. The project leader, Dr. Yutaka Nomura, ordered me to develop the medical diagnostic logic for ECG data through the Fisher's LDF and QDF. Although I had hoped to become a mathematical researcher when I was a senior student in high school, I failed the entrance examination of graduate school at Kyoto University because I spent much more time pursuing the activities of the swimming club in the university. Although I did not become a mathematical researcher, I started research with ME. The research I conducted from 1971 to 1974 using Fisher's LDF and QDF was inferior to his experimental decision tree logic. Initially, I believed that my statistical ability was poor. However, I soon realized that Fisher's assumption was too strict for medical diagnosis. I proposed the earth model (Shinmura, 1984)[1] for

---

[1] See the references in Chap. 1.

medical diagnosis instead of Fisher's assumption. Therefore, this experience gave me the motivation to develop the Theory. Shinmura et al. (1973, 1974) proposed a spectrum diagnosis using Bayesian theory that was the first trial for the Theory. However, logistic regression was more suitable for the earth model.

Shimizu et al. (1975) requested me to analyze photochemical air pollution data by Hayashi quantification theory, and this became my first paper. Dr. Takaichirou Suzuki, leader of the Epidemiology Group, provided me with several themes for many types of cancers (Shinmura et al. 1983).

In 1975, I met Prof. Akihiko Miyake from the Nihon Medical School at the workshop organized by Dr. Shigekoto Kaihara, Professor Emeritus of the Medical School of Tokyo University. Miyake and Shinmura (1976) studied the relationship between population and sample error rate in Fisher's LDF. Next, Miyake and Shinmura (1979) developed an OLDF based on the MNM criterion by a heuristic approach. Shinmura and Miyake (1979) discriminated CPD data with collinearities. After we revised a paper two or three times, a statistical journal rejected our paper. However, Miyake and Shinmura (1980) was accepted by Japanese Society for Medical and Biological Engineering (JSMBE). Former editors who judged OLDF based on the MNM criterion overestimated the validation sample, and Fisher's LDF did not overestimate the sample because Fisher's LDF was derived from the normal distribution without examination of real data. I was deeply disappointed that many statisticians disliked real data review and started their research from a normal distribution because it was very comfortable for them without the examination of real data (lotus eating). However, I could not develop a second trial of the Theory because of poor computer power and a defect in the heuristic approach.

Shinmura et al. (1987) analyzed the specific substance mycobacterium (SSM, commonly known as Maruyama vaccine). From 270,000 patients, we categorized 152,289 cancer patients into four postoperative groups. Those patients that were administered SSM within one year after surgery were divided into four groups every three months at the start of the SSM administration. We assumed that SSM is only water without side effects, and this was the null hypothesis. The survival time for the first group was longer than for the fourth group from nine months to 12 months after surgery and the null hypothesis was rejected.

In 1994, Prof. Kazunori Yamaguchi and Michiko Watanabe strongly recommended me to apply for the position at Seikei University. After organizing the 9th Symposium of JSCS in SCSK at Ryogoku near Ryogoku Kokugikan in March 1995, I became a professor at the Economic Department in April of the same year. Dr. Tokuhide Doi presented a long-term care insurance system that employed a decision tree method as advised by me. (Doctor Kaihara planned this system as an advisor to the Ministry of Health and Welfare, and I advised Dr. Doi to use the decision tree.)

In 1997, Prof. Tomoyuki Tarumi advised me to obtain a doctorate degree in science at his graduate school. Without examining the previous research, I developed IP-OLDF and LP-OLDF that discriminated Iris data, CPD data, and 115 random number datasets. IP-OLDF found two relevant facts about the Theory. Therefore, we confirmed the MNM criterion was essential for the discriminant analysis and complete the Theory in 2015. The Theory is useful for the gene datasets as same as the ordinary datasets. Redears can download all my research from researchmap and Theory from research gate.

https://www.researchgate.net/profile/Shuichi_Shinmura
http://researchmap.jp/read0049917/?lang=english

Musashinoshi, Japan                                                                 Shuichi Shinmura

# Acknowledgments

I wish to thank all researchers who contributed to this book: Linus Schrage, Kevin Cunningham, Hitoshi Ichikawa, John Sall, Noriki Inoue, Kyoko Takenaka, Masaichi Okada, Naohiro Masukawa, Aki Ishii, Ian B. Jeffery, Tomoyuki Tarumi, Yutaka Tanaka, Kazunori Yamaguchi, Michiko Watanabe, Yasuo Ohashi, Akihiko Miyake, Shigekoto Kaihara, Akira Ooshima, Takaichirou Suzuki, Tadahiko Shimizu, Tatuo Aonuma, Kunio Tanabe, Hiroshi Yanai, Toji Makino, Jirou Kondou, Hiroshi Takamori, Hidenori Morimura, Atsuhiro Hayashi, Iebun Yun, Hirotaka Nakayama, Mika Satou, Masahiro Mizuta, Souichirou Moridaira, Yutaka Nomura, Naoji Tuda.

I am grateful for my families, in particular, the legacy of my late father, who supported the research: Otojirou Shinmura, Reiko Shinmura, Makiko Shinmura, Hideki Shinmura, and Kana Shinmura.

I would like to thank Editage (www.editage.jp) for English language editing.

# Contents

1 **New Theory of Discriminant Analysis** .................... 1
   1.1 Introduction ........................................ 1
      1.1.1 Theory Theme .................................. 1
      1.1.2 Five Problems ................................. 3
         1.1.2.1 Problem 1 ............................ 3
         1.1.2.2 Problem 2 ............................ 4
         1.1.2.3 Problem 3 ............................ 4
         1.1.2.4 Problem 4 ............................ 5
         1.1.2.5 Problem 5 ............................ 5
         1.1.2.6 Summary .............................. 6
   1.2 Motivation for Our Research ......................... 6
      1.2.1 Contribution by Fisher ........................ 6
      1.2.2 Defect of Fisher's Assumption for Medical Diagnosis .... 7
      1.2.3 Research Outlook .............................. 8
      1.2.4 Method 1 and Problem 4 ........................ 9
   1.3 Discriminant Functions .............................. 10
      1.3.1 Statistical Discriminant Functions ............. 10
      1.3.2 Before and After SVM .......................... 11
      1.3.3 IP-OLDF and Four New Facts of Discriminant Analysis ... 13
      1.3.4 Revised IP-OLDF, Revised LP-OLDF, and Revised IPLP-OLDF .... 16
   1.4 Unresolved Problem (Problem 1) ...................... 17
      1.4.1 Perception Gap of Problem 1 ................... 17
      1.4.2 Student Data .................................. 18
   1.5 LSD Discrimination (Problem 2) ...................... 20
      1.5.1 Importance of This Problem .................... 20
      1.5.2 Pass/Fail Determination ....................... 21
      1.5.3 Discrimination by Four Testlets ............... 22

|   | 1.6 | Generalized Inverse Matrices (Problem 3) | 24 |
|---|---|---|---|
|   | 1.7 | *K*-Fold Cross-Validation (Problem 4) | 25 |
|   |   | 1.7.1   100-Fold Cross-Validation | 25 |
|   |   | 1.7.2   LOO and *K*-Fold Cross-Validation | 25 |
|   | 1.8 | Matroska Feature-Selection Method (Problem 5) | 28 |
|   | 1.9 | Summary | 29 |
|   | References | | 31 |
| **2** | **Iris Data and Fisher's Assumption** | | **37** |
|   | 2.1 | Introduction | 37 |
|   |   | 2.1.1   Evaluation of Iris Data | 37 |
|   |   | 2.1.2   100-Fold Cross-Validation for Small Sample (Method 1) | 38 |
|   | 2.2 | Iris Data | 39 |
|   |   | 2.2.1   Data Outlook | 39 |
|   |   | 2.2.2   Model Selection by Regression Analysis | 41 |
|   | 2.3 | Comparison of Seven LDFs | 42 |
|   |   | 2.3.1   Comparison of MNM and Eight NMs | 42 |
|   |   | 2.3.2   Comparison of Seven Discriminant Coefficient | 43 |
|   |   | 2.3.3   LINGO Program 1: Six MP-Based LDFs for Original Data | 44 |
|   | 2.4 | 100-Fold Cross-Validation for Small Sample Method (Method 1) | 48 |
|   |   | 2.4.1   Four Trials to Obtain Validation Sample | 48 |
|   |   | 2.4.1.1   Generate Training and Validation Samples by Random Number | 48 |
|   |   | 2.4.1.2   20,000 Normal Random Sampling | 50 |
|   |   | 2.4.1.3   20,000 Resampling Samples | 50 |
|   |   | 2.4.1.4   *K*-Fold Cross-Validation for Small Sample Method | 50 |
|   |   | 2.4.2   Best Model Comparison | 50 |
|   |   | 2.4.3   Comparison of Discriminant Coefficient | 52 |
|   | 2.5 | Summary | 53 |
|   | References | | 54 |
| **3** | **Cephalo-Pelvic Disproportion Data with Collinearities** | | **57** |
|   | 3.1 | Introduction | 57 |
|   | 3.2 | CPD Data | 58 |
|   |   | 3.2.1   Collinearities | 58 |
|   |   | 3.2.2   How to Find Linear Relationships in Collinearities | 60 |
|   |   | 3.2.3   Comparison Between MNM and Eight NMs | 64 |
|   |   | 3.2.4   Comparison of 95 % CI of Discriminant Coefficient | 66 |
|   | 3.3 | 100-Fold Cross-Validation | 66 |
|   |   | 3.3.1   Best Model | 66 |
|   |   | 3.3.2   95 % CI of Discriminant Coefficient | 68 |

|   |   | 3.4 | Trial to Remove Collinearity | 69 |
|---|---|---|---|---|
|   |   |   | 3.4.1 Examination by PCA (Alternative 2) | 69 |
|   |   |   | 3.4.2 Third Alternative Approach | 71 |
|   |   | 3.5 | Summary | 78 |
|   |   | References | | 79 |
| 4 | **Student Data and Problem 1** | | | 81 |
|   |   | 4.1 | Introduction | 81 |
|   |   | 4.2 | Student Data | 82 |
|   |   |   | 4.2.1 Data Outlook | 82 |
|   |   |   | 4.2.2 Different LDFs | 84 |
|   |   |   | 4.2.3 Comparison of Seven LDFs | 85 |
|   |   |   | 4.2.4 K-Best Option | 87 |
|   |   |   | 4.2.5 Evaluation by Regression Analysis | 87 |
|   |   | 4.3 | 100-Fold Cross-Validation of Student Data | 89 |
|   |   |   | 4.3.1 Best Model | 89 |
|   |   |   | 4.3.2 Comparison of Coefficients by LINGO Program 1 and Program 2 | 91 |
|   |   | 4.4 | Student Linearly Separable Data | 93 |
|   |   |   | 4.4.1 Comparison of MNM and Nine "Diff1s" | 93 |
|   |   |   | 4.4.2 Best Model | 93 |
|   |   |   | 4.4.3 95 % CI of Discriminant Coefficient | 95 |
|   |   | 4.5 | Summary | 97 |
|   |   | References | | 97 |
| 5 | **Pass/Fail Determination Using Examination Scores** | | | 99 |
|   |   | 5.1 | Introduction | 99 |
|   |   | 5.2 | Pass/Fail Determination Using Examination Scores Data in 2012 | 100 |
|   |   | 5.3 | Pass/Fail Determination by Examination Scores (50 % Level in 2012) | 102 |
|   |   |   | 5.3.1 MNM and Nine NMs | 102 |
|   |   |   | 5.3.2 Error Rate Means (M1 and M2) | 103 |
|   |   |   | 5.3.3 95 % CI of Discriminant Coefficients | 103 |
|   |   | 5.4 | Pass/Fail Determination by Examination Scores (90 % Level in 2012) | 106 |
|   |   |   | 5.4.1 MNM and Nine NMs | 106 |
|   |   |   | 5.4.2 Error Rate Means (M1 and M2) | 107 |
|   |   |   | 5.4.3 95 % CI of Discriminant Coefficient | 108 |
|   |   | 5.5 | Pass/Fail Determination by Examination Scores (10 % Level in 2012) | 109 |
|   |   |   | 5.5.1 MNM and Nine NMs | 110 |
|   |   |   | 5.5.2 Error Rate Means (M1 and M2) | 110 |
|   |   |   | 5.5.3 95 % CI of Discriminant Coefficients | 111 |
|   |   | 5.6 | Summary | 112 |
|   |   | References | | 113 |

| | | | |
|---|---|---|---|
| **6** | **Best Model for Swiss Banknote Data** | | 117 |
| | 6.1 Introduction | | 117 |
| | 6.2 Swiss Banknote Data | | 118 |
| | | 6.2.1 Data Outlook | 118 |
| | | 6.2.2 Comparison of Seven LDF for Original Data | 120 |
| | 6.3 100-Fold Cross-Validation for Small Sample Method | | 122 |
| | | 6.3.1 Best Model Comparison | 122 |
| | | 6.3.2 95 % CI of Discriminant Coefficient | 124 |
| | |     6.3.2.1 Consideration of 27 Models | 124 |
| | |     6.3.2.2 Revised IP-OLDF | 126 |
| | |     6.3.2.3 Hard-Margin SVM (H-SVM) and Other LDFs | 128 |
| | 6.4 Explanation 1 for Swiss Banknote Data | | 130 |
| | | 6.4.1 Matroska in Linearly Separable Data | 130 |
| | | 6.4.2 Explanation 1 of Method 2 by Swiss Banknote Data | 132 |
| | 6.5 Summary | | 135 |
| | References | | 136 |
| **7** | **Japanese-Automobile Data** | | 139 |
| | 7.1 Introduction | | 139 |
| | 7.2 Japanese-Automobile Data | | 140 |
| | | 7.2.1 Data Outlook | 140 |
| | | 7.2.2 Comparison of Nine Discriminant Functions for Non-LSD | 143 |
| | | 7.2.3 Consideration of Statistical Analysis | 144 |
| | 7.3 100-Fold Cross-Validation (Method 1) | | 147 |
| | | 7.3.1 Comparison of Best Model | 147 |
| | | 7.3.2 95 % CI of Coefficients by Six MP-Based LDFs | 148 |
| | |     7.3.2.1 Revised IP-OLDF Versus H-SVM | 148 |
| | |     7.3.2.2 Revised IPLP-OLDF, Revised LP-OLDF, and other LDFs | 151 |
| | | 7.3.3 95 % CI of Coefficients by Fisher's LDF and Logistic Regression | 151 |
| | 7.4 Matroska Feature-Selection Method (Method 2) | | 152 |
| | | 7.4.1 Feature-Selection by Revised IP-OLDF | 152 |
| | | 7.4.2 Coefficient of H-SVM and SVM4 | 155 |
| | 7.5 Summary | | 158 |
| | References | | 158 |
| | Bibliography | | 160 |

## 8 Matroska Feature-Selection Method for Microarray Dataset (Method 2) ........................................... 163
- 8.1 Introduction ........................................ 163
- 8.2 Matroska Feature-Selection Method (Method 2) .............. 165
  - 8.2.1 Short Story to Establish Method 2 ................. 165
  - 8.2.2 Explanation of Method 2 by Alon et al. Dataset ......... 167
    - 8.2.2.1 Feature-Selection by Eight LDFs ............. 167
    - 8.2.2.2 Results of Alon et al. Dataset Using the LINGO Program ................. 168
  - 8.2.3 Summary of Six Microarray Datasets in 2016 .......... 169
  - 8.2.4 Summary of Six Datasets in 2015 .................. 173
- 8.3 Results of the Golub et al. Dataset ....................... 173
  - 8.3.1 Outlook of Method 2 by the LINGO Program 3 ......... 173
  - 8.3.2 First Trial to Find the Basic Gene Sets ............... 177
  - 8.3.3 Another BGS in the Fifth SM ..................... 179
- 8.4 How to Analyze the First BGS .......................... 181
- 8.5 Statistical Analysis of SM1 ............................. 183
  - 8.5.1 One-Way ANOVA ............................. 183
  - 8.5.2 Cluster Analysis .............................. 184
  - 8.5.3 PCA ....................................... 185
- 8.6 Summary .......................................... 186
- References ............................................. 186

## 9 LINGO Program 2 of Method 1 ............................ 191
- 9.1 Introduction ........................................ 191
- 9.2 Natural (Mathematical) Notation by LINGO ................. 192
- 9.3 Iris Data in Excel .................................... 194
- 9.4 Six LDFs by LINGO .................................. 196
- 9.5 Discrimination of Iris Data by LINGO ..................... 199
- 9.6 How to Generate Resampling Samples and Prepare Data in Excel File ........................................ 200
- 9.7 Set Model by LINGO ................................. 202

**Index** ................................................. 205

# Symbols

## Statistical Discriminant Functions by JMP

| | |
|---|---|
| JMP | Statistical software supported by the JMP division of SAS Institute, Japan |
| JMP script | JMP script solves Fisher's LDF and logistic regression by Method 1 |
| LDF | linear discriminant functions such as Fisher's LDF, logistic regression, two OLDFs, three revised OLDFs, and three SVMs |
| Fisher's LDF | Fisher's linear discriminant function under Fisher's assumption |
| Logist | Logistic regression; in the table, "Logist" is often used |
| QDF* | Quadratic discriminant function |
| RDA* | Regularized discriminant analysis |
| * | QDF and RDA discriminate ordinary data in this book |

## Mathematical Programming (MP) by LINGO and What's Best!

| | |
|---|---|
| What's Best! | Excel add-in solver |
| LINGO | MP solver that can solve LP, IP, QP, NLP, and stochastic programming |
| LINGO Program 1 | LINGO that solves the original data by six MP-based LDFs explained in 2.3.3 |
| LINGO Program 2 | LINGO that solves six MP-based LDFs by Method 1 explained in Chap. 9 |
| LINGO Program 3 | LINGO that solves six MP-based LDFs by Method 2 |
| LP | Linear Programming develops Revisel LP-OLDF |

| | |
|---|---|
| IP | Integer Programming develops Revised IP-OLDF |
| QP | Quadratic Programming develops three SVMs |
| NLP | Nonlinear Programming defines Fisher's LDF |

## MP-based LDFs

| | |
|---|---|
| SVM | Support vector machine |
| H-SVM | Hard-margin SVM |
| S-SVM | Soft-margin SVM |
| SVM4$^{**}$ | S-SVM for penalty c = 10000 |
| SVM1$^{**}$ | S-SVM for penalty c = 1 |
| ** | Because there is no rule to decide a proper "c", we compare results by SVM4 and SVM1 |

## OLDF: Optimal LDF

| | |
|---|---|
| LSD | Linearly separable data, MNM of which is zero; LSD includes several linearly separable models (or subspaces) |
| Matroska | In gene analysis, we call all linearly separable space and subspaces as Matroska |
| Big Matroska | the microarray dataset is LSD and includes smaller Matroska in it by monotonic decrease of MNM |
| SM | small Matroska found by LINGO program 3 not explained in this book |
| BGS | basic gene set or subspace that is the smallest Matroska in each SM |
| NM | Number of misclassifications |
| MNM | Minimum NM |
| CP | Convex polyhedron; the interior point of CP has unique NM and discriminates same cases defined by IP-OLDF, not Revised IP-OLDF |
| OCP | Optimal CP; the interior point of OCP has unique MNM |
| IP-OLDF | OLDF-based MNM criterion using IP; if data are not general position, IP-OLDF may not find true MNM |
| Revised IP-OLDF | It finds the interior point of OCP and solves Problem 1 |
| LP-OLDF | OLDF using LP; one of L1-norm LDF |
| Revised LP-OLDF | One of L1-norm LDF; Although it is faster than other MP-based LDFs, it is weak for Problem 1 |
| Revised IPLP-OLDF | A mixture model of Revised LP-OLDF in the first step and Revised IP-OLDF in the second step |

# DATA

| | |
|---|---|
| Data | n cases by p-independent variables |
| $x_i$ | ith p-independent variable vector (for i = 1,...,n) |
| $y_i$ | Object variable; $y_i$=1 for class 1 and $y_i$=-1 for class 2 |
| $H_i(b)$ | $H_i(b) = y_i \times (^t x_i \times b + 1)$ is a linear hyperplane (for i = 1, ..., n) and divide p-dimensional coefficient space into finite CP (two half-planes such as $H_i(b) < 0$ and $H_i(b) > 0$.) |
| $H_i(b) < 0$ | Minus half-plane of $H_i(b)$: If $H_i(b_k) < 0$, $H_i(b_k) = y_i \times (^t x_i \times b_k + 1) = y_i \times (^t b_k \times x_i + 1) < 0$ and case $x_i$ is misclassified. If interior point $b_k$ is located in h-minus half-plane, NM = h. This LDF misclassifies the same h–cases |

## Ordinary or common Data

| | |
|---|---|
| Iris data | Fisher evaluates Fisher's LDF by these data |
| CPD data | Cephalo-pelvic disproportion data with collinearities |
| Student data | Pass/fail determination using student attribute |
| LSD | Linearly separable data that include linearly separable models in it. In gene analysis, we call LSD and linearly separable models are Matroskas |
| Swiss banknote data | IP-OLDF finds these data are LSD; We explain Problem 2 and 5 |
| Test data | Pass/fail determination using examination scores; these datasets are LSD, and we explain a trivial LDF |
| Japanese-automobile data | LSD; we explain Problem 3 and 5 |
| The datasets | Six microarray datasets |

## Theory and Method

| | |
|---|---|
| Theory | New theory of discriminant analysis after R. Fisher |
| Method 1 | 100-fold cross-validation for small sample method |
| Method 2 | Matroska feature-selection method for microarray datasets |
| M1 | The mean of error rate in the training sample |
| M2 | The mean of error rate in the validation sample |
| Best model | M2 of best model is minimum among all possible models of each LDF |
| LOO procedure | A leave-one-out model selection procedure |
| The best model | The model with minimum M2 instead of LOO by Method 1 |
| Diff1 | The difference defined as (NM of nine discriminant functions—MNM) |
| Diff | The difference defined as (M2—M1) |

| | |
|---|---|
| M1Diff | The difference defined as (M1 of nine discriminant functions-M1 of Revised IP-OLDF) |
| M2Diff | The difference defined as (M2 of nine discriminant functions-M2 of Revised IP-OLDF) |

## Five Problems of Discriminant Analysis

| | |
|---|---|
| Problem 1 | All LDFs, with the exception of Revised IP-OLDF, cannot discriminate the cases on the discriminant hyperplane. NMs of these LDFs may not be correct |
| Problem 2 | All LDFs, with the exception of H-SVM and Revised IP-OLDF, cannot recognize LDF theoretically; Although Revised LP-OLDF and Revised IPLP-OLDF can often discriminate LSD, we never discuss in Chap. 8 because of this reason |
| Problem 3 | The defect of the generalized inverse matrices technique; QDF misclassifies all cases as other classes for a particular case. Adding a small random noise to the constant values solves Problem 3 |
| Problem 4 | Fisher never formulated an equation for the standard error of the error rate and discriminant coefficient. Method 1 offers 95 % confidence interval (CI) for the error rate and coefficient. |
| Problem 5 | For more than ten years, many researchers have struggled to analyze the microarray dataset that is LSD. Only Revised IP-OLDF can make feature-selection naturally. I develop the Matroska feature-selection method (Method 2) that finds a surprising structure of the microarray dataset where such structure is the disjoint unions of several small linearly separable subspaces (small Matroska, SMs). Now we can analyze each SM very quickly. Student linearly separable, Swiss banknote, and Japanese-automobile data show the natural feature-selection of Revised IP-OLDF. Therefore, I recommend that researchers of feature-selection methods, such as LASSO, evaluate and compare their theory through these datasets in Chaps. 4, 6–8. I omit the results of the pass/fail determination using examination scores that consist only four variables |

# Chapter 1
# New Theory of Discriminant Analysis

## 1.1 Introduction

### 1.1.1 Theory Theme

This book introduces a new theory of discriminant analysis (hereafter, "the Theory") after R. Fisher. This chapter explains how to solve the five serious problems of discriminant analysis. To the best of my knowledge, this is the first book that compares eight linear discriminant functions (LDFs) using several different types of data. These eight LDFs are as follows: Fisher's LDF (Fisher 1936, 1956), logistic regression (Cox 1958), hard-margin SVM (H-SVM) (Vapnik 1995), two soft-margin SVMs (S-SVMs) such as SVM4 (penalty $c$ = 10,000) and SVM1 (penalty $c$ = 1), and three optimal LDFs (OLDFs). At first, I develop an OLDF based on a minimum number of misclassifications (minimum NM (MNM)) criterion using integer programming (IP-OLDF) and an OLDF using linear programming (LP-OLDF) (Shinmura 2000b, 2003, 2004, 2005, 2007). However, because I find the defect of IP-OLDF, I develop three revised OLDFs such as Revised IP-OLDF (Shinmura 2010a, 2011a), Revised LP-OLDF, and Revised IPLP-OLDF (Shinmura 2010b, 2014b). Iris data in Chap. 2 are critical test data because Fisher evaluates Fisher's LDF with these data (Anderson 1945). Cephalo-pelvic Disproportion (CPD) data (Miyake and Shinmura 1980) in Chap. 3 are medical data with three collinearities. Although Student data in Chap. 4 employ a small data sample (Shinmura 2010a), we can understand Problem 1 because the data are not general positions. The 18 pass/fail determinations using examination scores in Chap. 5 are linearly separable data (LSD). None of the LDFs, with the exception of H-SVM and Revised IP-OLDF, can discriminate LSD theoretically. I demonstrate that 18 error rates of Fisher's LDF and the quadratic discriminant function (QDF) are very high (Shinmura 2011b); nevertheless, these data are LSD. Moreover, seven LDFs, with the exception of Fisher's LDF, become trivial LDF (Shinmura 2015b). Swiss banknote data (Flury and Rieduyl 1988) in Chap. 6 and

© Springer Science+Business Media Singapore 2016
S. Shinmura, *New Theory of Discriminant Analysis After R. Fisher*,
DOI 10.1007/978-981-10-2164-0_1

Japanese-automobile data (Shinmura 2016c) in Chap. 7 are also LSD. Although I develop a Matroska feature-selection method for microarray dataset (Method 2), it is difficult for us to understand the meaning of Method 2 if we do not know LSD discrimination very well. I call LSD as big Matroska. As same as big Matroska includes several small Matroska, the microarray dataset (the datasets) includes several linearly separable subspaces (small Matroska (SM)) in it (the largest Matroska). Therefore, I explain this idea using common data in Chaps. 6 and 7. When I discriminate the datasets, only Revised IP-OLDF can select features naturally and finds the surprising structure of the datasets (Shinmura 2015e–s, 2016b).

Moreover, I develop a 100-fold cross-validation for small sample method (Method 1) (Shinmura 2010a, 2013, 2014c) instead of the leave-one-out (LOO) procedure (Lachenbruch and Mickey 1968). We can obtain two error rate means, $M1$ and $M2$, from the training and validation samples, respectively, and propose a simple model selection procedure to select the best model with minimum $M2$. The best model of Revised IP-OLDF is better than the seven other $M2$s from the previous data except for the Iris data.

We cannot discriminate cases on the discriminant hyperplane (Problem 1). Only Revised IP-OLDF can solve Problem 1. Moreover, only H-SVM and Revised IP-OLDF can discriminate LSD theoretically (Problem 2). Problem 3 is the defect of the generalized inverse matrix technique and QDF of misclassifying all cases to another class for a particular case. I solve Problem 3. Fisher never formulated an equation for the standard errors(SEs) of the error rate and discriminant coefficient (Problem 4). The Method 1 offers the 95 % confidence interval (CI) of the error rate and coefficient. For more than ten years, many researchers have struggled to analyze the *dataset* that is LSD (Problem 5). Only Revised IP-OLDF can make feature-selection naturally. The Method 2 finds the surprising structure of the dataset that is the disjoint unions of several small gene subspaces (SMs) that are linearly separable models. If we can repair the specific genes found by Method 2, we might overcome cancer diseases. Now, we can analyze each SM very quickly. We call the linearly separable model in gene analysis, "Matroska." If the datasets are LSD, the full model is the largest Matroska that contains all smaller Matroska in it. We already know that the smallest Matroska (the basic gene set or subspace (BGS)) can describe the Matroska structure completely by monotonic decrease of MNM. On the other hand, LASSO (Buhlmann and Geer 2011; Simon et al. 2013) attempts to make the feature-selection similar to Method 2. This book offers useful datasets and results for LASSO researchers from the following perspective:

1. Can LDF obtained by LASSO discriminate three different types of LSD such as Swiss banknote data, Japanese-automobile data, and six microarray datasets exactly?
2. Can LDF obtained by LASSO find the Matroska structure correctly and list all BGSs?

If LASSO cannot find SMs or BGS in the dataset, it cannot explain the data structure.

## 1.1.2 Five Problems

The Theory discusses only binary or two-class, class 1 or class 2, discrimination by eight LDFs such as Revised IP-OLDF, Revised LP-OLDF, Revised IPLP-OLDF, H-SVM, SVM4, SVM1, Fisher's LDF, and logistic regression. The values of class 1 and class 2 are 1 and −1, respectively. We consider these values as object variable of discriminant analysis and regression analysis. Let $f(\mathbf{x})$ be LDF and $f(\mathbf{x_i})$ be a discriminant score for $\mathbf{x_i}$. Although there are many difficult statistics in discriminant analysis, we should focus on the discriminant rule that is quite direct: If $y_i \times f(\mathbf{x_i}) > 0$, $\mathbf{x_i}$ is classified into class 1/class 2 correctly. If $y_i \times f(\mathbf{x_i}) < 0$, $\mathbf{x_i}$ is misclassified. If $y_i \times f(\mathbf{x_i}) = 0$, we cannot discriminate $\mathbf{x_i}$ correctly. This understanding is most important for discriminant analysis. There are five serious problems hidden in this simplistic scenario (Shinmura 2014a, 2015c, d).

### 1.1.2.1 Problem 1

We cannot adequately discriminate between cases where $\mathbf{x_i}$ lies on the discriminant hyperplane ($f(\mathbf{x_i}) = 0$). The Student data in Chap. 4 show this fact clearly. Thus far, this has been an unresolved problem. However, most researchers classify these cases into class 1 without logical reason. They misunderstand the discriminant rule as follows: If $f(\mathbf{x_i}) \geq 0$, $\mathbf{x_i}$ is classified into class 1 correctly. If $f(\mathbf{x_i}) < 0$, $\mathbf{x_i}$ is classified into class 2 properly. There are two mistakes in their rule. The first mistake is to classify the cases on the discriminant hyperplane to class 1 without logical explanation. The second mistake is we cannot determine the cases with positive discriminant score as classified into class 1 and those with a negative value as classified into class 2 a priori because the data determine this, not researchers. Other statisticians propose determining Problem 1 randomly (i.e., akin to throwing dice) because statistics is the study of probabilities. If users would know of this claim, they might be surprised and disappointed in discriminant analysis. In particular, medical doctors might be upset because they do not gamble with medical diagnoses, given that they attempt to seriously discriminate cases based on the discriminant hyperplane. Most statistical researchers are lack of this fact of medical diagnosis. If we consider pass/fail determination using the scores of four tests where the passing mark is 50 points, we can obtain trivial LDF such as $f = T1 + T2 + T3 + T4 - 50$. If $f \geq 0$, a given student has passed the examination. On the other hand, if $f < 0$, the student has failed the examination. Because we can describe the discriminant rule by (independent) variables clearly, we can correctly include such student on the discriminant hyperplane in the passing class. We have ignored this unresolved problem until now. The proposed Revised IP-OLDF based on MNM can treat Problem 1 appropriately (Shinmura 2010a). Indeed, with the exception of Revised IP-OLDF, no LDFs can correctly count the number of misclassifications (NMs). Therefore, we must count the number of cases where $f(\mathbf{x_i}) = 0$ and display this number "$h$" alongside NM of all LDFs in the output. We must estimate a true NM that might increase to (NM + $h$). After showing many examples of Problem 1, some statisticians claim that the probability of cases on the discriminant hyperplane is zero without a theoretical reason. They erroneously believe that we discriminate data on a continuous space.

### 1.1.2.2 Problem 2

Only H-SVM and Revised IP-OLDF can recognize LSD theoretically.[1] Other LDFs might not discriminate LSD exactly. When IP-OLDF discriminates Swiss banknote data in Chap. 6, I find that these data are LSD. In addition, Japanese-automobile data are LSD in Chap. 7. Through both data, I explain the Matroska feature-selection method (Method 2) in Chap. 8. We can obtain examination scores easily, and these datasets are also LSDs. Moreover, there is trivial LSD. However, several LDFs cannot determine pass/fail using examination scores correctly (Shinmura 2015b). In particular, the error rates of Fisher's LDF and QDF are very high. Table 1.4 lists all the 18 error rates of Fisher's LDF and QDF that are not zero in the pass/fail determinations from 2010 to 2012. This fact suggests that review the discriminant analysis of past important research because error rates may decrease. In medical diagnosis, researchers gave up their researches, error rates of which were over ten percent. However, Revised IP-OLDF may tell them error rates are zero. Moreover, discriminant functions that cannot discriminate LSD correctly are not helpful for gene analysis.

### 1.1.2.3 Problem 3

If the variance–covariance matrix is singular, Fisher's LDF and QDF cannot calculate it because inverse matrices do not exist. Because JMP (Sall et al. 2004) adopted the generalized inverse matrix technique, I had believed that Fisher's LDF and QDF could calculate generalized inverse matrix without problems. When I discriminated math examination scores among 56 examination data from the National Center for University Entrance Examinations (NCUEE), QDF and a regularized discriminant analysis (RDA) (Friedman 1989) misclassified all students in the passing class as the failing class. If we exchange class 1 and class 2, QDF and RDA misclassified all students in the failing class as the passing class decided by JMP specification. When QDF caused serious problems with problematic data, JMP switched QDF to RDA automatically. After three years of surveys, I found that RDA and QDF do not work correctly for a particular case where the values of the variables that belong to one class have a constant value because all the students in the passing class answered the particular question correctly. If users can select

---

[1]Empirically, Revised LP-OLD can discriminate LSD correctly. However, it is very weak for Problem 1. Logistic regression and SVM4 discriminate LSD correctly for many examinations. Fisher's LDF, QDF, and SVM1 are severe for LSD discriminations. I recommend researchers review their old researches using these three discriminant functions.

appropriate options for a modified RDA developed for this particular case, RDA works better than the QDF listed in Table 1.5, which is explained by the results of the Japanese-automobile data. However, JMP does not currently offer a modified QDF. Therefore, I judged this was the defect of generalized inverse matrix. If we add slight random noise to the constant value, QDF can discriminate the data exactly. Because it is the basic statistical knowledge for us, the data varied and I trust the quality of JMP; I need three years to find the reason. Problem 3 has provided a warning for our statistical understanding data always change.

### 1.1.2.4 Problem 4

Some statisticians erroneously believe that discriminant analysis is the inferential statistical method that is similar to regression analysis. However, Fisher never formulated an equation of SEs for discriminant coefficients or error rates. Nonetheless, if we use the indicator $y_i$ of mathematical programming-based linear discriminant functions (MP-based LDFs) in Eq. (1.7) as the object variable and analyze the data by regression analysis, the obtained regression coefficients are proportional to the coefficients of Fisher's LDF by the plug-in rule1. Therefore, we use a model selection procedure, such as stepwise procedures, and all possible combination models (Goodnight 1978) with statistics such as AIC, BIC, and Cp of regression analysis. In this book, I propose Method 1 and the new model selection procedure such as the best model. I set $k = 100$ and select the model with minimum $M2$ as the best model; this is a very direct and powerful model selection procedure compared with LOO. First, we select the best model in each LDF. Next, we select the model with minimum $M2$ among six MP-based LDFs as the final best model. We claim that the final best model has generalization ability. Moreover, we obtain the 95 % CI of the discriminant coefficient. Although we could demonstrate in 2010 that the best model was useful (Shinmura 2010a), I could not explain the useful meaning of the 95 % CI of the discriminant coefficient before 2014. However, if we divide all coefficients by the LDF intercept and set the intercept to one, six MP-based LDFs and logistic regression become trivial LDFs, and only Fisher's LDF is far from trivial (Shinmura 2015b). Moreover, I can explain the useful meaning of the 95 % CI of Swiss banknote and Japanese-automobile data (Shinmura 2016a, c) more precisely.

### 1.1.2.5 Problem 5

For more than ten years, many researchers have struggled to analyze the datasets (Problem 5). However, to the best of my knowledge, there has been no research on LSD discrimination thus far. I examine five different types of LSDs, such as Swiss banknote data, pass/fail determination of 18 examination data, Japanese-automobile data, student linearly separable data and six microarray datasets. When I discriminate the datasets, most of the coefficients of Revised IP-OLDF become zero. Only

Revised IP-OLDF can select features naturally and finds the surprising structure of the datasets. The datasets are Alon et al. (1999), Chiaretti et al. (2004), Golub et al. (1999), Shipp et al. (2002), Singh et al. (2002), and Tian et al. (2003). Jeffery et al. (2006) analyzed these datasets and upload these datasets on their HP.[2] Ishii et al. (2014) analyzed these datasets by principal component analysis (PCA). I find the Matroska structure in the datasets, with MNM of zero. The Method 2 can reduce the high-dimensional gene space into several small Matroskas (SMs) (Shinmura 2015e–s, 2016a). We can analyze these SMs by ordinary statistical methods such as $t$ test, one-way ANOVA, cluster analysis, and PCA. Because there has been no research on LSD discrimination thus far (to the best of our knowledge), many researchers have struggled and have not obtained good results. I explain Method 2 with the results of Swiss banknote data in Chap. 6 and Japanese-automobile data in Chap. 7 because Revised IP-OLDF can select variables naturally for ordinary data.

#### 1.1.2.6 Summary

Revised IP-OLDF solves Problems 1, 2, and 5. Problem 3 is the defect of the generalized inverse matrix technique, and QDF now causes Problem 3. If we add slight random noise to the constant value, we can solve Problem 3 easily. I propose Method 1 and compare two statistical LDFs by JMP script and six MP-based LDFs by the LINGO Program 2 (Schrage 2006) using six different types of data. Through many results, I can confirm that Method 1 solves Problem 4 using a computer-intensive approach. Problem 5 is the complex analysis of microarray datasets. Only Revised IP-OLDF can make feature-selection of the datasets naturally and find the datasets that consist of several disjoint unions of SMs. We can analyze each SM in the dataset easily because each SM is a small gene subspace. It is quite strange three SVMs cannot select feature naturally.

## 1.2 Motivation for Our Research

### 1.2.1 *Contribution by Fisher*

Fisher described Fisher's LDF using variance–covariance matrices and founded the statistical discriminant theory. He assumed that two classes (or groups) have the same variance–covariance matrices, and two means are different (Fisher's assumption). However, because Fisher's assumption is too strict for actual data, QDF was defined as two classes having different variance–covariance matrices. This fact indicates that statisticians are aware that there exist data that do not satisfy Fisher's assumption. Moreover, multiclass discrimination that uses the

---

[2]http://www.bioinf.ucd.ie/people/ian/.

Mahalanobis distance has been proposed. In the quality control, Taguchi and Jugular (2002) considered that one class (the normal state) has a variance–covariance matrix, and another class (the uncontrolled state) consists of only one case. They discriminated data through multiclass discrimination and claimed that the typical uncontrolled case is far from the normal state with large Mahalanobis distance. Their claim is similar to the "earth model" in medical diagnosis (Shinmura 1984). Because statistical software packages easily implement these discriminant functions based on variance–covariance matrices, we apply discriminant analysis to many applications in science, technology, and industry, such as medical diagnosis, pattern recognition, and various ratings. However, real data rarely satisfy Fisher's assumptions. Therefore, it is well known that logistic regression is better than Fisher's LDF and QDF because it does not assume a particular theoretical distribution, such as a normal distribution. It is very strange and unfortunate for us that there is no discussion on this matter by researchers and users of logistic regression.

### 1.2.2 Defect of Fisher's Assumption for Medical Diagnosis

After graduating from Kyoto University in 1971, I became a member of the project that developed the automatic diagnostic system for electrocardiogram (ECG) data from 1971 to 1974. A project leader who was a medical doctor requested me to discriminate over ten[3] abnormal symptoms from normal symptom using Fisher's LDF and QDF. Our four years of research were inferior to medical doctor's experimental decision tree logic. First, I believed that my results using Fisher's LDF and QDF were inferior to decision tree logic results because my knowledge and experience was poor. Later, I realized that Fisher's assumption was not adequate for medical diagnosis. I summarized two reasons for my failure, both of which are described below. On the other hand, there is no actual test for Fisher's assumption. I demonstrate that NM of Fisher's LDF is close to MNM in Iris data. We can use this trend instead of the test statistics of Fisher's hypothesis.

*First Reason*: In medical diagnosis, typical cases in abnormal symptoms are far from the discriminant hyperplane. I explained medical diagnosis as the "earth model" where the normal symptom is the land, abnormal symptoms are the mountains, and the discriminant hyperplanes are horizon. The Mahalanobis–Taguchi strategy is similar to the earth model. This claim violates Fisher's assumption. In a statistical concept, we understand that typical cases in both classes are two averages of two normal distributions. Therefore, I believed that the discriminant functions based on the variance–covariance matrices are not adequate for medical diagnosis and developed a spectrum diagnostic method (Shinmura et al. 1973, 1974). I knew that logistic regression is remarkably successful in medical diagnosis and understood that it is superior to the spectrum diagnostic method.

---

[3] I cannot recollect the exact number of abnormal symptoms.

Currently, Japanese medical researchers discriminate data by logistic regression instead of Fisher's LDF and QDF. I regret that as researchers and users of logistic regression, they did not discuss my claim.

*Second Reason*: There are many cases close to the discriminant hyperplane. I concluded that Fisher's LDF and QDF are fragile for the discrimination of particular data, such as pass/fail determination using examination scores (Shinmura 2011b) and the rating of bonds, stocks, and estates in addition to medical data. These data also have the characteristic feature of having many cases close to the discriminant hyperplane. None of the LDFs, with the exception of Revised IP-OLDF, can discriminate the cases on the discriminant hyperplane correctly (Problem 1). Recently, because I could not access medical data for our research, I used pass/fail determination with examination scores instead of medical data.

## 1.2.3 Research Outlook

After 1975, I discriminated many data using Fisher's LDF, QDF, logistic regression, multiclass discrimination using Mahalanobis distance, decision tree logic (or partitioning), and the quantification theory developed by Dr. Hayashi (Shimizu et al. 1975; Nomura and Shinmura 1978; Shinmura et al. 1983). Through these studies, I found Problems 1 and 4 (Shinmura 2014a, 2015c, d). In 1973, we developed the spectrum diagnostic method using Bayesian theory. However, logistic regression was more sophisticated than the spectrum diagnostic method. Next, we developed OLDF based on the MNM criterion (Miyake and Shinmura 1979, 1980; Shinmura and Miyake 1979), which is a heuristic approach. Because Warmack and Gonzalez (1973) compared several discriminant functions, their research encouraged our research. We were not able to develop the research because we had low computer power and because of the defect of the heuristic approach.

Starting in 1997, I developed IP-OLDF (Shinmura 1998; 2000a, b; Shinmura and Tarumi 2000). Because I defined IP-OLDF in the discriminant coefficient spaces, I found two important facts of discriminant analysis. The first is OCP. The second is "the monotonic decrease of MNM." However, there was a serious defect in IP-OLDF using Student data that are not general positions. If data are not general positions, IP-OLDF might not search for the vertex of a true OCP. This defect means that the obtained MNM might not be true MNM, and Problem 1 caused this defect. In 2007, Revised IP-OLDF solved the defect because it can find the interior point of true OCP and avoid Problem 1. Therefore, I could solve Problem 1 completely. Until 2007, I was not able to evaluate eight LDFs using validation samples because our research data were small samples.

After 2007, I developed Method 1. Through this breakthrough, I was able to solve Problem 4 and ended the basic research. Revised IP-OLDF solves Problems 1 and 2. Although I can evaluate eight LDFs by $M2$, I cannot explain the useful meaning of the 95 % CI of discriminant coefficients. After 2010, I started applied research on LSD discrimination. I found that Problem 3 is the defect of the generalized inverse matrix technique by the pass/fail determination that uses examination scores (Shinmura 2011b). With regard to IP-OLDF, I set the intercept of IP-OLDF to one and was able to obtain two important facts such as OCP and monotonic decrease of MNM. Therefore, I divided all coefficients by the intercept and set the intercept to one. Through the second breakthrough, seven LDFs, with the exception of Fisher's LDF, became trivial LDF by the pass/fail determination that uses examination scores, and I was able to explain the useful meaning of the coefficient of Revised IP-OLDF using Swiss banknote data. Therefore, I have solved the four problems and can confirm the end of our research. However, when I discriminated Shipp et al. dataset in October 2015, I found that Revised IP-OLDF can make feature-selection naturally and can solve Problem 5 quickly.

## 1.2.4 Method 1 and Problem 4

If we set "$k = 100$" in the Method 1, we can obtain 100 LDFs and 100 error rates from the training and validation samples. From the 100 LDFs, we obtain the 95 % CI of discriminant coefficients. From the 100 error rates, we obtain the 95 % CI of error rates and two means of error rates, $M1$ and $M2$, from the training and validation samples. We consider the model with minimum $M2$ among all possible combination models to be the best model. This standard is a direct and powerful model selection procedure compared with the LOO procedure.

We should distinguish such computer-intensive approaches from traditional inferential statistics with the SE equation based on normal distribution. Statisticians without computer power established inferential statistics manually. Today, we can utilize the power of a computer with statistical and MP solvers, such as JMP and LINGO. I developed the Method 1 (Program 2) of Fisher's LDF and logistic regression with the JMP script supported by the JMP division of SAS Institute Japan. In addition, I developed Method 1 for six MP-based LDFs with LINGO. Those are Revised IP-OLDF, Revised IPLP-OLDF, Revised LP-OLDF, H-SVM, SVM4, and SVM1. I explain the LINGO Program 2 in Chap. 9. Those researchers who want to analyze their research data can obtain the 95 % CI for the error rate and discriminant coefficients. These statistics provide precise and deterministic judgment on model selection procedure compared with the LOO procedure. To this point, I cannot validate and evaluate Revised IP-OLDF with seven other LDFs because I only have small original data and no validation samples. Researchers with small samples can validate and assess their research data with Method 1 and the best model.

Miyake and Shinmura (1976) discussed "error rates of linear discriminant function" by the traditional approach. On the other hand, Konishi and Honda (1992) discussed "error rate estimation using the bootstrap method." Their computer-intensive approaches are not traditional inferential statistics and do not offer the 95 % CI of the error rates and coefficients for individual data. Although logistic regression outputs the 95 % CI of the coefficient through maximum likelihood proposed by R. Fisher, this is also a computer-intensive approach. On the other hand, we can select the best model and 95 % CI of the error rates and coefficients for six LDFs by the Method 1 and the best model. Many researchers who want to discriminate small samples have the Philosopher's Stone.

## 1.3 Discriminant Functions

I compare two statistical LDFs by JMP and six MP-based LDFs by LINGO. I omit a kernel SVM because it is a nonlinear discriminant function. However, I evaluate QDF and RDA with eight LDFs only for the original six different data, with the exception of the datasets. Next, I compare two statistical LDFs and six MP-based LDFs for resampling samples if the data are LSD. If the data are not LSD, we cannot discriminate the data by H-SVM because it causes error for non-LSD.

### 1.3.1 Statistical Discriminant Functions

Fisher defined Fisher's LDF by maximization of the variance ratio (between/within classes) in Eq. (1.1). Nonlinear programming (NLP) can solve this equation.

$$\text{MIN} = {}^t\mathbf{b}(\mathbf{m}_1 - \mathbf{m}_2){}^t(\mathbf{m}_1 - \mathbf{m}_2)\mathbf{b}/{}^t\mathbf{b}\Sigma\mathbf{b} \tag{1.1}$$

If we accept Fisher's assumption, the same LDF is obtained in Eq. (1.2) by another plug-in rule2. This equation defines Fisher's LDF explicitly, whereas Eq. (1.1) defines LDF implicitly. Therefore, statistical software packages adopt this equation. Some statisticians erroneously believe that discriminant analysis is inferential statistics, similar to regression analysis. Discriminant analysis is not traditional inferential statistics based on the normal distribution because there are no SEs for the discriminant coefficients and error rates (Problem 4). Therefore, Lachenbruch and Mickey proposed the LOO procedure for selecting a good discriminant model, as indicated in Table 1.6.

$$\text{Fisher's LDF}: f(\mathbf{x}) = {}^t\{\mathbf{x} - (\mathbf{m}_1 + \mathbf{m}_2)/2\}\ \Sigma^{-1}(\mathbf{m}_1 - \mathbf{m}_2) \tag{1.2}$$

Most real data do not satisfy Fisher's assumption. When the variance–covariance matrices of two classes are not the same ($\Sigma_1 \neq \Sigma_2$), the QDF defined in Eq. (1.3) can be used. This fact is critical for us. Previous statisticians have known that most real data do not satisfy Fisher's assumption. We use the Mahalanobis distance in Eq. (1.4) for the discrimination of multiclasses. The Mahalanobis–Taguchi method of quality control is one of the applications.

## 1.3 Discriminant Functions

$$\text{QDF}: f(\mathbf{x}) = {}^t\mathbf{x}(\Sigma_2^{-1} - \Sigma_1^{-1})\mathbf{x}/2 + \left({}^t\mathbf{m}_1\Sigma_1^{-1} - {}^t\mathbf{m}_2\Sigma_2^{-1}\right)\mathbf{x} + c \quad (1.3)$$

$$D = \text{SQRT}\left({}^t(\mathbf{x}-\mathbf{m})\Sigma^{-1}(\mathbf{x}-\mathbf{m})\right) \quad (1.4)$$

We use Fisher's LDF and QDF in many areas, but cannot calculate whether some variables remain constant. There are three cases. First, some variables that belong to both classes are the same constant. Second, some variables that belong to both classes are different, but constant. Third, some variables that belong to one class are constant. Most statistical software packages exclude all variables in these three cases. On the other hand, JMP enhances QDF using the generalized inverse matrix technique. Therefore, QDF can treat the first and second cases correctly, but cannot manage the third case properly (Problem 3).

Recently, the logistic regression in Eq. (1.5) has been used instead of Fisher's LDF and QDF for two reasons. First, it is well known that the error rate of logistic regression is often less than that of Fisher's LDF and QDF because it is derived from real data, instead of some normal distribution free from reality. Let "$p$" be the probability of belonging to a class of diseases. If the value of some variable is increasing/decreasing, "$p$" increases from zero (normal class) to one (abnormal class). This representation is very useful in medical diagnosis, as well as for ratings in real estates and bonds. On the contrary, Fisher's LDF assumes that cases close to the average of the diseases are representative cases of the diseases' class. Medical doctors never permit this claim. Although the maximum-likelihood procedure calculates SE of the logistic coefficient, we should distinguish the computer-intensive approach from the traditional inferential statistics based on the theoretical distribution induced manually. Firth (1993) indicated that the SE of a logistic coefficient becomes large and the convergence calculation becomes unstable for LSD. If I observe the following points: (1) I can find NM = 0 by changing the discriminant hyperplane on ROC, (2) MNM = 0, (3) SEs become large, and (4) the convergence calculation becomes unstable, I can determine that logistic regression can recognize LSD. I confirm that logistic regression can almost recognize LSD by this tedious work:

$$\text{Log}\left(p/(1-p)\right) = f(\mathbf{x}) \quad (1.5)$$

### 1.3.2 Before and After SVM

There are many types of research on MP-based discriminant analysis. Glover (1990) defined many linear programming (LP) discriminant models. Rubin (1997) proposed MP-based discriminant functions using IP. Stam (1997) summarized Lp-norm discriminant methods in 1997 and answered the question, "Why have statisticians rarely used Lp-norm methods?" He provided four reasons: communication, promotion, and terminology; software availability; the relative accuracy of

Lp-norm classification methods: ad hoc studies; and the accuracy of Lp-norm classification methods: decision theoretic justification. Although each of these reasons is true, they are not important. The most important reason is that there is no comparison between these methods with statistical discriminant functions because discriminant analysis was established by Fisher before MP approaches. There are two types of MP applications. The first is modeling by MP, such as for portfolio selection (Markowitz 1959) that is similar to S-SVM. The second is catch-up modeling, such as for regression and discriminant analyses (Schrage 1991). Therefore, the latter type should be compared with the preceding results. To the best of my knowledge, no statisticians use Lp-norm methods because there is no research that indicates that MP-based methods are superior to statistical methods. Liitschwager and Wang (1978) defined a model based on the MNM criterion shown in Eq. (1.6) that is very close to Revised IP-OLDF. Although there are several mistakes in their model, the most important is a restriction on the discriminant coefficients. If they could have confirmed their model with an IP solver, they might have found the defect of their model quickly. I should set the intercept to one. There is no need to set the other ($k-1$) coefficients in the range [$-1, 1$].

$$\text{MIN } p_1 r_1 M^{-1} \Sigma_{(i=1,\ldots,M)} P_i + p_2 r_2 N^{-1} \Sigma_{(j=1,\ldots,N)} Q_j$$
$$\text{st}$$
$$c_1 X_{i1} + c_2 X_{i2} + \cdots + c_k X_{ik} \leq b + CP_i, \ i = 1, 2, \ldots, M$$
$$c_1 Y_{j1} + c_2 Y_{j2} + \cdots + c_k Y_{jk} \geq b - CQ_j, \ j = 1, 2, \ldots, N \quad (1.6)$$
$$-1 + 2 D_r \leq c_r \leq 1 - 2 E_r, \ r = 1, 2, \ldots, k$$
$$\Sigma_{(r=1,\ldots k)} D_r + \Sigma_{(r=1,\ldots,k)} E_r = 1$$

$P_i, Q_i, D_r, E_r$: 0/1 decision variable
$c_1, c_2, \ldots, c_k, b$: free variables
$C$: large constant, such as 10,000
$p_1, p_2$: prior probability
$r_1, r_2$: risk by misclassification
$M, N$: number of cases of two groups
$k$: number of independent variables

Vapnik (1995) proposed three different SVM models. H-SVM indicates the discrimination of LSD clearly. IP-OLDF confirms that Swiss banknote data are LSD and realize the importance of Problem 2 by H-SVM. This is defined as the maximization of the distance of the "support vector (SV)" in order to obtain "good generalization ability," which is similar to "not overestimating the validation data in statistics." It is redefined to minimize (1/"distance of SV") in Eq. (1.7). A quadratic programming (QP) solves it that can analyze the only LSD, not overlapping data. This restriction might ignore the LSD investigation. Statisticians erroneously believe that LSD discrimination is very easy. In statistics, there was no technical term for LSD before H-SVM. However, the condition "MNM = 0" is the same as being linearly separable. Note that "NM = 0" does not imply that the data are linearly separable. It is unfortunate that there has been no research into linear separability (Problems 2 and 5).

$$\text{MIN} = ||\mathbf{b}||^2/2; \quad y_i \times ({}^t\mathbf{x_i}b + b_0) \geq 1; \tag{1.7}$$

**b**: *p*-discriminant coefficients. $b_0$: H-SVM intercept
$y_i = 1/-1$ for $\mathbf{x_i} \in$ class 1/class 2. $\mathbf{x_i}$: *p*-variables (independent variables)

Real data are rarely linearly separable. Therefore, S-SVM is defined in Eq. (1.8). S-SVM permits certain cases that are not discriminated by SV ($y_i \times ({}^t\mathbf{x_i}b + b_0) < 1$). The second objective is to minimize the summation of distances of misclassified cases ($\Sigma e_i$) from SV. These two objects are combined by defining some "penalty *c*." The Markowitz portfolio model that minimizes risk and maximizes return is the same as S-SVM. However, return is incorporated as a constraint, and the objective function minimizes only risk. The decision maker selects a solution on the efficiency frontier. On the contrary, S-SVM does not have a rule for determining c correctly; nevertheless, it can be solved by an optimization solver. Therefore, I compare two S-SVMs, such as SVM4 (*c* = 10,000) and SVM1 (*c* = 1). In many trials, NM of SVM4 is less than NM of SVM1.

$$\text{MIN} = ||\mathbf{b}||^2/2 + c \times \Sigma e_i; \quad y_i \times ({}^t\mathbf{x_i}\mathbf{b} + b_0) \geq 1 - e_i \tag{1.8}$$

*c*: penalty c for combining two objectives. $e_i$: nonnegative value

### 1.3.3 IP-OLDF and Four New Facts of Discriminant Analysis

Miyake and Shinmura (1979, 1980) and Shinmura and Miyake (1979) developed a heuristic algorithm of OLDF based on the MNM criterion. This algorithm solves the five-variable model of CDP data that consists of two groups with 19 variables explained in Chap. 3. I introduced SAS into Japan in 1978 and three technical reports on the generalized inverse matrix, sweep operator (Goodnight 1978), and SAS regression applications (Sall 1981) related to this research. I introduced LINDO to Japan in 1983. MP (Schrage 1991) formulated several regression models, e.g., QP can solve least-squares regression, and LP can solve least absolute value (LAV) regression. Without a survey of previous research, I formulated IP-OLDF in Eq. (1.9). This notation is defined in the p-dimensional coefficient space because I set the intercept to one. In pattern recognition, the intercept is a free variable. In this case, the model is defined in the (*p* + 1) coefficient space, and we cannot elicit the same deep knowledge as with IP-OLDF. This difference is crucial. I can consider IP-OLDF in both the *p*-dimensional data and coefficient spaces. We can apparently understand the relationship between NM and LDF. The linear equation $H_i(\mathbf{b}) = y_i \times ({}^t\mathbf{x}_i\mathbf{b} + 1) = 0$ divides the $p$-dimensional coefficient space into positive and negative half-planes ($H_i(\mathbf{b}) > 0$, $H_i(\mathbf{b}) < 0$). If $\mathbf{b}_j$ is in the positive half-plane, $f_j(\mathbf{x}) = y_i \times ({}^t\mathbf{b}_j\mathbf{x} + 1)$ discriminates $\mathbf{x}_i$ correctly because $f_j(\mathbf{x}_i) = y_i \times ({}^t\mathbf{b}_j\mathbf{x}_i + 1) = y_i \times ({}^t\mathbf{x}_i\mathbf{b}_j + 1) > 0$. On the contrary, if $\mathbf{b}_j$ is included in the negative half-plane, $f_j(\mathbf{x})$ cannot discriminate $\mathbf{x}_i$ correctly because $f_j(\mathbf{x}_i) = y_i \times ({}^t\mathbf{b}_j\mathbf{x}_i + 1) = y_i \times ({}^t\mathbf{x}_i\mathbf{b}_j + 1) < 0$. Then, the linear equation $H_i(\mathbf{b})$ can divide the coefficient space into a finite number of CPs. Each CP interior point has a unique NM that is equal to the number of negative half-planes. I define OCP as that value for which NM is equivalent to MNM. If $\mathbf{x}_i$ is classified correctly, $e_i = 0$ and $H_i(\mathbf{b}) \geq 0$ in Eq. (1.9). If there are p-cases on $f(\mathbf{x}_i) = 0$, we can obtain the exact MNM. However, if there are over $(p + 1)$ cases on $f(\mathbf{x}_i) = 0$, this causes Problem 1. If $\mathbf{X}_i$ is misclassified, $e_i = 1$ and $H_i(\mathbf{b}) \geq -10,000$. This means that IP-OLDF selects the discriminant hyperplane $H_i(\mathbf{b}) = 0$ for correctly classified cases and $H_i(\mathbf{b}) = -10,000$ for misclassified cases according to a 0/1 decision variable. IP-OLDF selects a vertex of the OCP with p-cases on $f(\mathbf{x}_i) = 0$. However, if the vertex consists of over $(p + 1)$ cases, MNM might not be correct. In addition to this defect, IP-OLDF must be solved for the three cases where the intercept is equal to 1, 0, and $-1$ because we cannot determine the sign of $y_i$ in advance. Combinations of $y_i = 1/-1$ for $\mathbf{x}_i \in$ class 1/class 2 are determined by the data, not the analyst. Many researchers do not know this important fact.

$$\text{MIN} = \Sigma e_i;\ H_i(\mathbf{b}) \geq -M \times e_i;$$
$$H_i(\mathbf{b}) = y_i \times ({}^t\mathbf{x}_i\mathbf{b} + 1) \qquad (1.9)$$
$$M : 10,000\ (\text{Big } M \text{ constant}).$$

Through IP-OLDF that uses Iris, CPD, and Swiss banknote data, I find four essential facts of discriminant analysis, as follows:

1. Because we define IP-OLDF in the discriminant coefficient space and set the intercept to one, we can understand the relationship between NM and the discriminant coefficient exactly. The interior points of specific CPs correspond to LDFs that misclassify the same cases. Therefore, the interior points have unique "NM." Because there are finite CPs, we should select the interior point of OCP, with NM of MNM. If we select the CP vertex or edge as LDF, this LDF is not free from Problem 1.
2. MNM decreases monotonously, such as $\text{MNM}_p \geq \text{MNM}_{(p+1)}$, because the $(p + 1)$-space includes the $p$-subspace. Because MNM of a full model has a minimum value, we cannot use MNM as the feature (model or variable) selection. If $\text{MNM}_k = 0$, all models, including these $k$-variables, are zero. This fact is critical for LSD discrimination and gene analysis. Swiss banknote data consist of six independent variables. I examine all possible models by IP-OLDF and find that MNM of the two-variable model ($X4, X6$) is zero. Therefore, 16 MNMs, including ($X4, X6$), are zero. The MNMs of 47 other models are not zero. This fact is essential for understanding the structure of microarray dataset in Chap. 8. We call the two-variable model ($X4, X6$) "BGS." Therefore, Swiss

## 1.3 Discriminant Functions

banknote data have one BGS, and we can understand Matroska structure of Swiss banknote data by (X4, X6) completely. This fact is very important for understanding of the datasets because there are numerous Matroskas in it.

3. Student data reveal the defect of IP-OLDF by Problem 1. IP-OLDF searches for the vertex of the correct OCP if the data are general positions and might not search for the vertex of the correct OCP if the data are not general positions. Therefore, I develop Revised IP-OLDF that searches from the interior point of true OCP directly.
4. If we compare NMs on both models selected by forward and backward stepwise procedures using CPD data, we observe that QDF are fragile for collinearity (Shinmura 2000a). Logistic regression is fragile for collinearity, also.

Let us consider the discrimination of three cases with two variables, as follows:

Class 1 : $x1 = (-1/18, -1/12)$
Class 2 : $x2 = (-1, 1/2)$, $x3 = (1/9, -1/3)$.

Equation (1.10) is the model for IP-OLDF:

$$\begin{aligned} \text{MIN} &= \Sigma e_i; \\ y_1 &\times \{-(1/18) \times b_1 - (1/12) \times b_2 + 1\} \geq -e_1; \\ y_2 &\times \{-b_1 + (1/2) \times b_2 + 1\} \geq -e_2; \\ y_3 &\times \{(1/9) \times b_1 - (1/3) \times b_2 + 1\} \geq -e_3; \end{aligned} \quad (1.10)$$

We consider the three linear Eqs. (1.11):

$$\begin{aligned} H_1 &= y_1 \times \{-(1/18) \times b_1 - (1/12) \times b_2 + 1\} = 0, \\ H_2 &= y_2 \times \{-b_1 + (1/2) \times b_2 + 1\} = 0, \\ H_3 &= y_3 \times \{(1/9) \times b_1 - (1/3) \times b_2 + 1\} = 0 \end{aligned} \quad (1.11)$$

The three linear equations divide the two-dimensional coefficient space into seven CPs, as shown in Fig. 1.1. The CP number is the NM of each LDF that is equal to the number of negative half-planes of $H_i(\mathbf{b})$ that surround CP. The interior point in the triangle is located in the three-plus hyperplanes, with NM of zero and MNM. Because two linear equations make three OCP vertexes, these data are general positions and free from Problem 1. NMs of three opposite CPs of OCP are one. Namely, NMs of adjacent CPs differ by one. Although we set the intercept to one, we must solve the three models as follows: intercept = 1, intercept = $-1$, and intercept = 0, because we cannot determine the sign of the discriminant score a priori. When we set the intercept to two, the graph shown in Fig. 1.1 is similarly enlarged to twice its size.

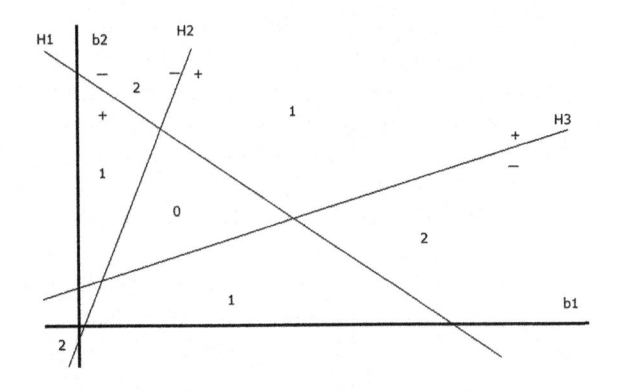

**Fig. 1.1** Relation between NM and discriminant coefficient

### 1.3.4 Revised IP-OLDF, Revised LP-OLDF, and Revised IPLP-OLDF

Revised IP-OLDF in Eq. (1.12) can find the exact MNM because it can directly find the OCP interior point. This means that there are no cases where $y_i \times (^t\mathbf{x}_i\mathbf{b} + b_0) = 0$. If $\mathbf{x}_i$ is discriminated correctly, $e_i = 0$ and $y_i \times (^t\mathbf{x}_i\mathbf{b} + b_0) \geq 1$. If $\mathbf{x}_i$ is misclassified, $e_i = 1$ and $y_i \times (^t\mathbf{x}_i\mathbf{b} + b_0) \geq -9999$. We expect that all misclassified cases will be extracted to second SVs, such as $y_i \times (^t\mathbf{x}_i\mathbf{b} + b_0) = -9999$. Therefore, the discriminant scores of the misclassified cases become a large negative less than $-1$, and there are no cases where $y_i \times (^t\mathbf{x}_i\mathbf{b} + b_0) = 0$. This means that $\mathbf{b}$ is an OCP interior point defined by IP-OLDF. Ibaraki and Muroga (1970) introduced the same model. However, they did not survey this model. If I had found this model first and started our survey with this model, I might not have established the Theory because I would have never struggled with IP-OLDF and obtained new facts.

$$\text{MIN} = \Sigma e_i; \quad y_i \times (^t\mathbf{x}_i\mathbf{b} + b_0) \geq 1 - M \times e_i; \\ b_0 : \text{free decision variable.} \tag{1.12}$$

If $e_i$ is a real nonnegative variable, the Eq. (1.12) utilizes Revised LP-OLDF, which is an L1-norm LDF. Its elapsed run time is faster than that of Revised IP-OLDF. If we select a large positive number, such as penalty $c$ for S-SVM, the result is almost the same as that given by Revised LP-OLDF because the role of the first term of the objective value in Eq. (1.7) decreases. Many trials realized that Revised LP-OLDF is fragile for Problem 1. Revised IPLP-OLDF is a combined model of Revised LP-OLDF and Revised IP-OLDF. In the first step, Revised LP-OLDF is applied for all cases, and $e_i$ is set to zero for cases that are discriminated correctly by Revised LP-OLDF. In the second phase, Revised IP-OLDF is used for the cases misclassified in the first step. Therefore, Revised IPLP-OLDF can

## 1.4 Unresolved Problem (Problem 1)

First, IP-OLDF reveals the following important properties:

Fact (1) Relationship between LDFs and NMs

IP-OLDF is defined in the discriminant coefficient spaces. Cases of $x_i$ that correspond to linear hyperplanes ($H_i$ (**b**) = $y_i \times$ (${}^t x_i \mathbf{b} + 1$) = 0) in the $p$-dimensional discriminant coefficient space divide the space into two half-planes: the positive ($H_i$ (**b**) > 0) and negative ($H_i$(**b**) < 0) half-planes. Therefore, the coefficient space is divided into a finite convex polyhedron by $H_i$(**b**). Interior point $\mathbf{b}_j$ of CP corresponds to LDF ($f_j(\mathbf{x}) = {}^t\mathbf{b}_j\mathbf{x} + 1$) in the data space that discriminates some cases appropriately and misclassifies others. This explanation means that each interior point $\mathbf{b}_j$ has a unique NM. OCP is defined as that with the MNM. Revised IP-OLDF finds the OCP interior point directly. Moreover, it solves the unresolved problem (Problem 1) because there are no cases on $f(\mathbf{x}_i) = 0$. If $\mathbf{b}_j$ is on a CP vertex or edge, however, the unresolved problem cannot be avoided because there are some cases on $f(\mathbf{x}_i) = 0$. In particular, I know that Revised LP-OLDF is weak for Problem 1 through many trials.

Fact (2) Monotonic decrease of MNM (MNM$_p$ ≥ MNM$_{(p+1)}$)

Let MNM$_p$ be the MNM of p-variables. Let MNM$_{(p+1)}$ be the MNM of the ($p + 1$)-variables formed by adding one variable to the original $p$-variables. MNM decreases monotonously (MNM$_p$ ≥ MNM$_{(p+1)}$) because OCP in the $p$-dimensional coefficient space is a subset of the ($p + 1$)-dimensional coefficient space. If MNM$_p$ = 0, all MNMs, including $p$-variables, are zero. Swiss banknote data consist of genuine and counterfeit bills with six variables. IP-OLDF finds that these data are LSD according to two variables (X4, X6). Therefore, 16 models, including these two variables, have MNMs = 0. Only Revised IP-OLDF can solve Problem 1 theoretically. Because (X4, X6) can explain all 16 models are linearly separable, BGS can explain the structure of Matroska in the microarray datasets completely.

### 1.4.1 Perception Gap of Problem 1

With regard to Problem 1, there are several misunderstandings. Most researchers treat cases $\mathbf{x}_i$ on $f(\mathbf{x}_i) = 0$ in class 1. There is no explanation for why this makes sense. Some statisticians explain that it is decided stochastically because statistics is the study of probability. This explanation seems theoretical initially, but it is nonsense for two reasons. Statistical software adopts the former decision rule because many papers and researchers adopt this rule. In medical diagnosis, medical doctors strive to determine patients close to the discriminant hyperplane. If they would know the second explanation, they might be deeply disappointed in discriminant analysis. To this point, all LDFs, such as Fisher's LDF, logistic regression, Revised LP-OLDF, Revised IPLP-OLDF, and S-SVM, have not been able to solve Problem 1 theoretically. IP-OLDF reveals that only interior points of the CP can solve Problem 1. IP-OLDF can find the vertex of the correct OCP if the data are general positions and stops the optimization from selecting p-cases on $f(\mathbf{x}_i) = 0$. However, IP-OLDF might not find the actual MNM if the data are not general positions, and it selects over $(p + 1)$ cases on $f(\mathbf{x}_i) = 0$. Revised IP-OLDF can find the OCP interior point directly. We cannot determine whether other LDFs select the CP interior point, edge, or vertex. We can confirm this fact in order to verify the number of cases $\mathbf{x}_i$ that satisfy $|f(\mathbf{x}_i)| \leq 10^{-6}$ if we consider $|f(\mathbf{x}_i)| \leq 10^{-6} = 0$ in the software. If this number is zero, this LDF selects the CP interior point exactly. If the number "h" is not zero, this LDF selects the CP vertex or edge, and the correct NM has a possibility of increasing to "h."

### 1.4.2 Student Data

Student data[4] are proper for us to discuss Problem 1. A total of 15 students ($y_i$ = "F") fail an examination, and 25 students ($y_i$ = "P") pass the examination, as indicated in Table 1.1. $X1$ is the number of study hours/day, and $X2$ is expenditure (10,000 yen)/month. In the case where IP-OLDF discriminates two classes by ($X1$, $X2$), the discriminant hyperplane of IP-OLDF is $X2 = 5$ solved by old version of "What's Best!." Eight students ($X2 > 5$) are discriminated against the failing group correctly; four students are on $X2 = 5$, and three students ($X2 < 5$) are misclassified into the passing group. On the other hand, 21 students ($X2 < 5$) are classified into the passing group correctly, and four students are on $X2 = 5$. Nevertheless, IP-OLDF cannot discriminate eight students in $X2 = 5$: it returns MNM = 3. However, Revised IP-OLDF by LINGO ver. 14 can find the three discriminant hyperplanes: $X2 = 0.006 \times X1 + 4.984$, $X2 = 0.25 \times X1 + 3.65$, and $X2 = 0.99 \times X1 + 212$.[5] Moreover, the correct MNM = 5. SVM4 is $X2 = X1 + 1$ and NM = 6. A student has the value (4, 5) on $f(\mathbf{x}_i) = 0$. Therefore, the true NM of SVM4 might be seven. Although these data are small, they are useful for evaluating LDFs, and they are easy for us to understand through scatter plots of two variables.

---

[4]These data were used for the description of four statistical books using SAS, SPSS, Statistica, and JMP. In this book, Chap. 4 includes the Student data.
[5]Because this problem is very nervous, the result may be different by the latest version of LINGO and other solvers.

# 1.4 Unresolved Problem (Problem 1)

**Table 1.1** Student data

| $y_i$ | F | F | F | F | F | F | F | F | F | F | F | F | F | F | F | P | P | P | P | P |
|---|---|---|---|---|---|---|---|---|---|---|---|---|---|---|---|---|---|---|---|---|
| X1 | 3 | 1 | 3 | 3 | 2 | 1 | 4 | 3 | 5 | 2 | 3 | 2 | 3 | 3 | 5 | 6 | 9 | 4 | 3 | 2 |
| X2 | 10 | 8 | 7 | 7 | 6 | 6 | 6 | 6 | 5 | 5 | 5 | 5 | 3 | 2 | 2 | 5 | 5 | 5 | 5 | 4 |
| $y_i$ | P | P | P | P | P | P | P | P | P | P | P | P | P | P | P | P | P | P | P | P |
| X1 | 5 | 12 | 4 | 10 | 7 | 5 | 7 | 3 | 7 | 7 | 7 | 6 | 3 | 6 | 6 | 8 | 5 | 10 | 9 | 5 |
| X2 | 4 | 4 | 4 | 4 | 4 | 4 | 3 | 3 | 3 | 3 | 3 | 3 | 3 | 3 | 3 | 3 | 3 | 2 | 2 | 2 |

## 1.5 LSD Discrimination (Problem 2)

### 1.5.1 Importance of This Problem

The purpose of discriminant analysis is to discriminate two classes correctly. For this reason, LSD discrimination is crucial because we can evaluate the results very clearly. If some LDFs cannot discriminate LSD correctly, such LDFs should not be used. It is very strange that there is no research on LSD discrimination (to the best of my knowledge). H-SVM implies LSD discrimination very clearly. However, it can be only applied to LSD. This restriction might be the reason for the lack of actual research on LSD until now. Some statisticians believe that OLDF based on the MNM criterion is bad LDF because it overfits the training samples, and its generalization ability might be wrong for validation samples without examination by real data. I confirm that Revised IP-OLDF does not overestimate and has better generalization ability than other LDFs including H-SVM through many trials.

IP-OLDF finds that Swiss banknote data are LSD with two variables ($X4$, $X6$), and MNMs of 16 models, including ($X4$, $X6$), are zero. To this point, nobody seems to have realized this fact. Moreover, we believe that it is difficult for us to find LSD from real data because we must discriminate all possible models by Revised IP-OLDF. However, we can easily obtain two types of research data: first, the pass/fail determination using examination scores. I explain the results in Chap. 5. Second, every real data are changed to LSD by enlarging the distance between the mean of the two classes in Chap. 4. Swiss banknote data consist of two types of bills: 100 genuine and 100 counterfeit. There are six variables: $X1$ is the bill length (mm); $X2$ and $X3$ are the width of the left and right edges (mm), respectively; $X4$ and $X5$ are the bottom and top margin widths (mm), respectively; and $X6$ is the length of the image diagonal (mm). I investigate a total of 63 ($=2^6 - 1$) models. According to Shinmura (2010a), of the 63 total models, 16 models, including the two variables ($X4$, $X6$), have MNMs of zero; thus, they are linearly separable models. The 47 models that remain are not linearly separable. These data are adequate regardless of whether LDFs can discriminate LSD correctly.

Table 1.2 lists four results. The upper right column (B) is the original data. The upper left column (A) is the data expanded to 1.25 times the average distance. The lower left (C) and right (D) columns are the data reduced to 0.75 and 0.5 times the average distance, respectively. Fisher's LDF is independent from inferential statistics. However, if we consider $y_i = 1/-1$ as the object value and analyze the data by regression analysis, the obtained regression coefficients are proportional to the discriminant coefficients of Fisher's LDF by the plug-in rule1. We can use the stepwise procedures formally. In the table, column "$p$" is the number of variables by the forward stepwise procedure. "Var" represents the selected variables. From $p = 1$ to $p = 6$, $X6$, $X4$, $X5$, $X3$, $X2$, and $X1$ are selected in this order by the forward stepwise procedure. In regression analysis, Mallow's Cp statistics and AIC are used for model selection. Usually, the model with minimum $|Cp - (p + 1)|$ and AIC is recommended. By this rule, Cp statistics selects the same full model. On the other

## 1.5 LSD Discrimination (Problem 2)

**Table 1.2** Swiss banknote data (Shinmura 2010a)

| | | A: the distance *1.25 | | | | B: original bank data | | | |
|---|---|---|---|---|---|---|---|---|---|
| Var. | p | Cp | AIC | MNM | LDF | Cp | AIC | MNM | LDF |
| 1–6 | 6 | **7.0** | −863 | 0 | 0 | **7.0** | −779 | 0 | 0 |
| 2–6 | 5 | 5.3 | −865 | 0 | 0 | 5.3 | **−781** | 0 | 0 |
| 3–6 | 4 | 10.5 | **−896** | 0 | 0 | 10.3 | −776 | 0 | 0 |
| 4–6 | 3 | 10.9 | −859 | 0 | 0 | 10.7 | −775 | 0 | 0 |
| 4, 6 | 2 | 118.8 | −779 | 0 | 0 | 107.0 | −699 | **0** | 3 |
| 6 | 1 | 313.9 | −679 | **0** | 1 | 292.0 | −604 | 2 | 2 |
| | | C: The distance * 0.75 | | | | D: The distance * 0.5 | | | |
| Var. | p | Cp | AIC | MNM | LDF | Cp | AIC | MNM | LDF |
| 1–6 | 6 | **7.0** | −676 | 1 | 2 | **7.0** | −543 | 5 | 12 |
| 2–6 | 5 | 5.3 | **−678** | 1 | 2 | 5.3 | **−545** | 6 | 12 |
| 3–6 | 4 | 9.8 | −673 | 1 | 1 | 8.9 | −541 | 7 | 13 |
| 4–6 | 3 | 10.1 | −673 | 1 | 2 | 8.8 | −541 | 8 | 14 |
| 4, 6 | 2 | 97.9 | −601 | 4 | 6 | 78.7 | −482 | 16 | 19 |
| 6 | 1 | 253.8 | −517 | 6 | 8 | 184.4 | −417 | 53 | 56 |

hand, AIC selects the four-variable model ($X3, X4, X5, X6$) in data "A." AIC selects the five-variable model ($X2, X3, X4, X5, X6$) in the other three data. This table indicates two important facts. We can easily obtain LSD from real data. I observe that the same result of the Swiss banknote data as the Student linearly separable data in Chap. 4. The second fact is as follows: "Cp and AIC" select the same models; nevertheless, the one-variable ($X6$) model is linearly separable in "A." Moreover, the two-variable ($X4, X6$) model is linearly separable in "B." The models selected by "Cp and AIC" are independent from linear separability. These facts show the defect of statistics based on the variance-covariance matrices. Some statisticians do not permit this result by the plug-in rule1. On the contrary, they consider Fisher's LDF as inferential statistics because it is derived from Fisher's assumption. However, they ignore Fisher never formulated the equation of SE of error rates and discriminant coefficients.

### 1.5.2 Pass/Fail Determination

The pass/fail determination using examination scores makes good research data because we can obtain such data quickly, and we can find trivial LDF as explained in Chap. 5. Our theoretical analysis started in 1997 and ended in 2010. Our applied research started in 2010 and ended in 2015. I negotiated with the National Center for University Entrance Examinations (NCUEE) and borrowed research data that consist of 105 examinations in 14 subjects over three years. I finished analyzing the data at the end of 2010 and obtained 630 error rates for Fisher's LDF, QDF, and Revised IP-OLDF. However, NCUEE requested me not to present the results in

March 2011. Therefore, I explain new research findings using my statistical examination results. I found the reason for the particular case of QDF and RDA (Problem 3) at the end of 2012. The course consists of one 90-min lecture per week for 15 weeks. In 2011, the course only ran for 11 weeks because of power shortages in Tokyo caused by the Fukushima nuclear accident. Approximately 130 students, mainly freshmen, attended the lectures. Midterm and final examinations consisted of 100 questions with ten choices. I found Problem 3 by the discrimination of 100-item scores. I discriminate two types of pass/fail determinations by 100-item scores and four testlet scores as variables. If the passing mark is 50 points, we can easily obtain trivial LDF ($f = T1 + T2 + T3 + T4 - 50$). If $f \geq 0$ or $f < 0$, the student passes or fails the examination, respectively. In this case, students on $f(\mathbf{x}_i) = 0$ pass the examination because their score is exactly 50. This fact indicates that there is no Problem 1 because the independent variables determine the discriminant rule clearly.

### 1.5.3 Discrimination by Four Testlets

Table 1.3 lists the discrimination of four testlet scores as variables for 10 % (from the third to the seventh column) and 90 % (after the eighth column) levels of the midterm examinations. I omit the results of the 50 % level from the table to save the space. In the table, "$p$" denotes the number of variables selected by the forward stepwise procedure. In 2010, T4, T2, T1, and T3 were entered in the model selected by the forward stepwise procedure. MNMs of Revised IP-OLDF and NM of logistic regression are zero in the full model, which means that the data are LSD in four variables. NMs of LDF and QDF are nine and two, respectively. This means that

**Table 1.3** NMs of four discriminant functions by forward stepwise in midterm examinations at 10 % (from the third to seventh column) and 90 % levels (after the eighth column)

| | $p$ | 10 % | | | | | 90 % | | | | |
|---|---|---|---|---|---|---|---|---|---|---|---|
| | | Var. | MNM | Logi. | LDF | QDF | Var. | MNM | Logi. | LDF | QDF |
| 2010 | 1 | T4 | 6 | 9 | 11 | 11 | T3 | 10 | 37 | 24 | 24 |
| | 2 | T2 | 2 | 6 | 11 | 9 | T4 | 5 | 10 | 20 | 11 |
| | 3 | T1 | 1 | 3 | 8 | 5 | T1 | *0* | *0* | 20 | 10 |
| | 4 | T3 | *0* | *0* | 9 | 2 | T2 | 0 | 0 | 20 | 11 |
| 2011 | 1 | T2 | 9 | 17 | 15 | 15 | T3 | 6 | 7 | 14 | 14 |
| | 2 | T4 | 4 | 9 | 11 | 9 | T4 | 1 | 1 | 14 | 6 |
| | 3 | T1 | *0* | *0* | 9 | 10 | T1 | *0* | *0* | 13 | 5 |
| | 4 | T3 | 0 | 0 | 9 | 11 | T2 | 0 | 0 | 14 | 9 |
| 2012 | 1 | T4 | 4 | 8 | 14 | 12 | T3 | 8 | 30 | 12 | 12 |
| | 2 | T2 | *0* | *0* | 11 | 9 | T1 | 5 | 12 | 9 | 9 |
| | 3 | T1 | 0 | 0 | 12 | 8 | T4 | 3 | 3 | 10 | 3 |
| | 4 | T3 | 0 | 0 | 12 | 1 | T2 | *0* | *0* | 11 | 3 |

## 1.5 LSD Discrimination (Problem 2)

**Table 1.4** Summary of error rates (%) of Fisher's LDF and QDF

|         |      | 10 % |     | 50 % |     | 90 % |      |
|---------|------|------|-----|------|-----|------|------|
|         |      | LDF  | QDF | LDF  | QDF | LDF  | QDF  |
| Midterm | 2010 | 7.5  | 1.7 | 2.5  | 5.0 | **16.7** | 9.2  |
|         | 2011 | 7.0  | 8.5 | **2.2** | 2.3 | 10.5 | 6.7  |
|         | 2012 | 9.9  | **0.8** | 4.9 | 4.8 | 13.6 | 7.1  |
| Final   | 2010 | 4.2  | 1.7 | 3.3  | 4.2 | 3.3  | **10.8** |
|         | 2011 | 11.9 | 2.9 | 2.9  | 3.6 | 3.6  | 8.6  |
|         | 2012 | 8.7  | 2.3 | 2.3  | 2.3 | 13.0 | 4.5  |

**Table 1.5** Discrimination of Japanese small and regular cars

| p | Var | t | LDF | QDF | MNM | $\lambda = \gamma = 0.8$ | 0.5 | 0.2 | 0.1 |
|---|-----|---|-----|-----|-----|------|-----|-----|-----|
| 1 | Emission | **11.37** | 2 | **0** | **0** | 2 | 1 | 1 | **0** |
| 2 | Price | 5.42 | 1 | **0** | **0** | 4 | 1 | 0 | 0 |
| 3 | Capacity | **8.93** | 1 | 29 | **0** | 3 | 1 | 0 | 0 |
| 4 | CO$_2$ | 4.27 | 1 | 29 | **0** | 4 | 1 | 0 | 0 |
| 5 | Fuel | −4.00 | **0** | 29 | **0** | 5 | 1 | 0 | 0 |
| 6 | Sales | −0.82 | 0 | 29 | **0** | 5 | 1 | 0 | 0 |

LDF and QDF cannot recognize LSD. In 2011, Revised IP-OLDF and logistic regression were able to acknowledge that the three-variable model (T2, T4, T1) is linearly separable. In 2012, the two-variable model (T4, T2) was linearly separable. T4 and T2 contain natural questions, and T1 and T3 consist of questions that are challenging to the students in the failing group. This suggests the possibility that pass/fail determination using Revised IP-OLDF can elicit the quality of test problems and understanding of pupils.

Table 1.4 lists the summary of the 18 error rates derived from NMs of Fisher's LDF and QDF for the linear separable model. The ranges of the 18 error rates of LDF and QDF are [2.2, 16.7 %] and [0.8, 10.8 %], respectively. The error rates of QDF are lower than those of LDF. At the 10 % level, the six error rates of LDF and QDF lie in the ranges [4.2, 11.9 %] and [0.8, 8.5 %], respectively. Clearly, the range at the 50 % level is less than that at the 10 and 90 % levels. Miyake and Shinmura (1976) followed Fisher's assumption and surveyed the relationship between population and sample error rates. One of our results suggests that the sample error rates of balanced sample sizes, such as the 50 % level, are close to the population error rates. The above results can confirm this result. Table 1.4 suggests a serious limitation of LDF and QDF based on the variance–covariance matrices. We can no longer trust the error rates of Fisher's LDF and QDF. To this point, this fact has not been discussed because there is slight research on using LSD (to the best of my knowledge). Now, I should evaluate discriminant functions using LSD because the results are very precise. In genome discrimination, researchers attempt to estimate the variance–covariance matrices using small sample sizes and large numbers of variables. These efforts might be meaningless and lead to incorrect results (Problem 5).

## 1.6 Generalized Inverse Matrices (Problem 3)

I confirm the particular cases found in the NCUEE examinations with our examinations. I found the reason for Problem 3 in November 2012. Three years were required because we never questioned the QDF algorithm and conducted our survey using the multivariate approach. I checked all variables by the $t$ test of two classes before abandoning the survey. I can explain the particular case by the discrimination of Japanese-automobile data.[6] Let us consider the discrimination of 29 regular and 15 small cars. Small cars have a unique Japanese specification. They are mainly sold as second cars or to women because they are cost-efficient. The emission rate of small and regular cars ranges from [0.657, 0.658] and [0.996, 3.456], respectively. The capacity (number of seats) of small and regular cars is 4 and [5, 8], respectively. Therefore, 48 models, including emission rate and capacity, are linearly separable. We call the emission rate and capacity "two basic gene sets or subspaces (BGS)" in Chap. 8. In addition, ($X4$, $X6$) of Swiss banknote data is one BGS that consists of two variables (or genes). We can understand the structure of the Japanese-automobile data by the emission rate and capacity in Chap. 7 and the structure of Swiss banknote data by ($X4$, $X6$) in Chap. 6. These facts are very important for us to understand Method 2.

Table 1.5 lists the forward stepwise result. First, "emission" is entered into the model because the $t$-value is high. MNM of Revised IP-OLDF and NMs of QDF are zero. Fisher's LDF cannot recognize LSD. Next, "price" is entered into the two-variable model, although the $t$-value of "price" is less than that of "capacity." In the third step, QDF misclassifies all 29 regular cars as small cars after "capacity" is entered into the three-variable model. This is because the capacity of small cars is four persons. It is critical for QDF and RDA to be the only ones affected by this particular case. Fisher's LDF and the $t$ test are not affected because we compute these statistics by the pooled variance of two classes. Modified RDA offers two options, such as $\lambda$ and $\gamma$. By the grid search, I find that $\lambda = \gamma = 0.1$ is better than the others. However, we must survey the grid search for all data. I expect the JMP division to display a guideline on how to select two parameters. As of 2015, JMP had not solved Problem 3 of QDF.

---

[6]These data are available in the paper on DEA (Table 2.1 in page 4. http://repository.seikei.ac.jp/dspace/handle/10928/402). Chap. 6 includes these data.

## 1.7 K-Fold Cross-Validation (Problem 4)

Usually, the LOO procedure is used for the model selection of discriminant analysis. I developed the Method 1 because it is stronger than the LOO procedure. Method 1 establishes a simple and powerful model selection procedure that selects the best model. Moreover, I can explain the meaning of 95 % CI of the discriminant coefficient.

### 1.7.1 100-Fold Cross-Validation

In regression analysis, we benefit from inferential statistics because SE of regression coefficients and model selection statistics, such as Cp, AIC, and BIC, are known a priori. On the other hand, there is no SE of discriminant coefficients and model selection statistics in discriminant analysis. Therefore, users of discriminant analysis and SVMs often use the LOO procedure. Let the sample size be "$n$." We use one case for validation and the other ($n - 1$) cases as the training samples. We evaluate n sets of training and validation samples. On the other hand, if we have a large sample size, we can use $k$-fold cross-validation by dividing the sample into $k$-subsamples. We can evaluate $k$-combinations of training and validation samples. On the other hand, bootstrap or resampling methods can be used with small sample sizes. In this research, we generate large sample sets by resampling and developing 100-fold cross-validation using these resampled data. Method 1 is as follows:

1. We copy 100 times the data from the original data using JMP.
2. We add a uniform random number as a new variable, sort the data in ascending order, and divide into 100 subsets.
3. We evaluate eight LDFs by Method 1 using these 100 subsets.

I analyze MP-based LDFs by LINGO, developed with the support of LINDO Systems Inc. I analyze logistic regression and Fisher's LDF by JMP, obtained with the assistance of the JMP division of SAS Japan. There is merit in using 100-fold cross-validation because we can easily calculate the 95 % CI of the discriminant coefficients and error rates. We can use the LOO procedure for model selection, but cannot obtain the 95 % CI. These differences are quite important for the analysis of small samples.

### 1.7.2 LOO and K-Fold Cross-Validation

Table 1.6 lists the results of the LOO procedure of Fisher's LDF and NMs of five LDFs in 2012 test data (10%). In the table, "Var" displays the suffix of four testlet scores named "$T$." There are only 11 models because I omit four one-variable

**Table 1.6** LOO and NMs in original test data

| No. | Var. | LOO | LDF | Logistic | MNM | SVM4 | SVM1 |
|-----|------|-----|-----|----------|-----|------|------|
| 1 | 1–4 | 14 | 12 | 0 | 0 | 0 | 0 |
| 2 | 1, 2, 4 | 13 | 12 | 0 | 0 | 0 | 2 |
| 3 | 2, 3, 4 | *11* | *11* | 0 | 0 | 0 | 0 |
| 4 | 1, 3, 4 | 15 | 15 | 2 | 2 | 3 | 3 |
| 5 | 1, 2, 3 | 16 | 16 | 6 | 4 | 6 | 6 |
| 6 | 2, 4 | *11* | *11* | *0* | *0* | *0* | 3 |
| 7 | 1, 4 | 16 | 16 | 6 | 3 | 6 | 6 |
| 8 | 3, 4 | 14 | 13 | 3 | 3 | 4 | 4 |
| 9 | 1, 2 | 18 | 17 | 12 | 7 | 7 | 7 |
| 10 | 2, 3 | 16 | 11 | 11 | 6 | 11 | 11 |
| 11 | 1, 3 | 22 | 21 | 15 | 7 | 10 | 10 |

models from the table. MNM of the two-variable model ($T2$, $T4$) in row No. 6 is zero in Table 1.3, as are those of the four-variable model ($T1$–$T4$), and the two three-variable models of ($T1$, $T2$, $T4$) and ($T2$, $T3$, $T4$). NMs of logistic regression and SVM4 are zero in these four models, but NMs of SVM1 are two and three in row No. 2 and 6, respectively. I often observe that S-SVM cannot recognize LSD when penalty c has a small value. The LOO procedure recommends the models in row No. 3 and 6 because their NMs are minimum.

Table 1.7 lists the results given by Revised IP-OLDF (RIP), SVM4, Fisher's LDF (LDF), and logistic regression (Logistic). I omit the results from SVM1, Revised LP-OLDF, and Revised IPLP-OLDF, because Revised IP-OLDF is better than three LDFs. The first column shows the same number as that in Table 1.6. After four linearly separable models, the ranges of seven models are displayed. The "$M1$" column denotes the error rate mean from the training sample. Revised IP-OLDF and logistic regression can recognize four linearly separable models. For SVM4, the only full model has an NM of zero. All $M1$s of Fisher's LDF are over 9.48 %. The "$M2$" column denotes the error rate mean from the validation sample. Because only two models (row nos 2 and 6) of Revised IP-OLDF have NMs of zero, we select this model as the best model. NMs of the other LDFs are greater than zero, and those of Fisher's LDF are over 9.91 %. We can conclude that Fisher's LDF is the worst of these four LDFs. Some statisticians believe that NMs of Revised IP-OLDF is less suitable for validation samples because it overfits the training samples. On the other hand, Fisher's LDF does not lead to overestimation because it assumes a normal distribution. Table 1.7 indicates that the presumption of "overestimation" is wrong. I might conclude that many real data do not obey Fisher's assumption. Discrimination based on an incorrect assumption will lead to incorrect results. "Diff" is the difference between $M2$ and $M1$. Some researchers believe that the small absolute value of "Diff" implies that there is no

1.7 K-Fold Cross-Validation (Problem 4)

**Table 1.7** Comparison of four functions

| RIP | M1 | M2 | Diff | Var | |
|---|---|---|---|---|---|
| 1 | 0 | 0.07 | 0.07 | 1, 2, 3, 4 | |
| 2 | 0 | 0 | 0 | 1, 2, 4 | |
| 3 | 0 | 0.03 | 0.03 | 2, 3, 4 | |
| 6 | 0 | 0 | 0 | 2, 4 | |
| 4, 5, 7–11 | [0.79, 4.94] | [0.03, 7.21] | [0.03, 2.39] | | |
| SVM4 | M1 | M2 | Diff | M1Diff | M2Diff |
| 1 | 0 | 0.81 | 0.81 | 0 | 0.74 |
| 2 | 0.73 | 1.62 | 0.90 | 0.73 | 1.62 |
| 3 | 0.13 | 0.96 | 0.83 | 0.13 | 0.93 |
| 6 | 0.77 | *1.70* | 0.93 | 0.77 | *1.70* |
| 4,5,7–11 | [1.65, 6.85] | [3.12, 8.02] | [0.66, 1.65] | [0.78, *2.33*] | [0.59, 1.36] |
| LDF | M1 | M2 | Diff | M1Diff | M2Diff |
| 1 | 9.64 | 10.54 | 0.90 | 9.64 | 10.47 |
| 2 | 9.89 | 10.55 | 0.66 | 9.89 | *10.55* |
| 3 | *9.48* | 10.09 | 0.61 | 9.48 | 10.06 |
| 6 | 9.54 | **9.91** | 0.37 | 9.54 | **9.91** |
| 4, 5, 7–11 | [10.81, 16.28] | [11.03, 16.48] | [0.16, 0.6] | [7.97, *11.34*] | [6.23, 9.61] |
| Logistic | M1 | M2 | Diff | M1Diff | M2Diff |
| 1 | 0 | 0.77 | 0.77 | 0 | 0.70 |
| 2 | 0 | 1.09 | 1.09 | 0 | 1.09 |
| 3 | 0 | 0.85 | 0.85 | 0 | 0.82 |
| 6 | 0 | **0.91** | 0.91 | 0 | **0.91** |
| 4, 5, 7–11 | [1.59, 7.65] | [2.83, 8.04] | [0.35, 1.34] | [0.8, *3.13*] | [0.39, *1.62*] |

overestimation. In this sense, Fisher's LDF is better than the other LDFs because all the values are less than 0.9 %. However, only the high values of $M1$ of Fisher's LDF lead to small values of "Diff." "$M1$Diff" is defined as the difference of ($M1$ of seven LDFs – $M1$ of Revised IP-OLDF) in the training samples, and "$M2$Diff" is the value of ($M2$ of seven LDFs – $M2$ of Revised IP-OLDF) in the validation samples. All values of "$M1$Diff and $M2$Diff" of SVM4, Fisher's LDF, and logistic regression are greater than zero. Fisher's LDF is not as good as the other LDFs with 100-fold cross-validation. Therefore, I should select the model of Revised IP-OLDF with the minimum value of $M2$ as the best model. Two models, such as ($T1$, $T2$, $T4$) and ($T2$, $T4$), are zero. In this case, I select the two-variable model ($T2$, $T4$) because of the principle of parsimony or Occam's razor. The values of "$M2$" for four LDFs are 0, 1.7, 9.91, and 0.91 %, respectively. This result implies that $M2$ of Fisher's LDF is 9.91 % higher than the best model of Revised IP-OLDF in the validation sample.

In 2014, these results were recalculated using LINGO version 14. The elapsed run times of Revised IP-OLDF and SVM4 were 3 min 54 s and 2 min 22 s, respectively. The elapsed run times of Fisher's LDF and logistic regression by JMP were 24 and 21 min, respectively. Reversals of CPU time have occurred for this time.

## 1.8 Matroska Feature-Selection Method (Problem 5)

In Chap. 8, the Method 2 discriminates the dataset quickly and shows the dataset consists of disjoint unions of several SMs that easily are analyzed by ordinary statistical methods. Let us consider that the dataset consists of two classes, such as cancer (50 cases) and normal (50 cases), with 10,000 genes. Our primary concern is to discriminate the two classes by 10,000 variables (genes). IP-OLDF finds two important facts in the discriminant coefficient space, as follows:

1. In the 100 linear hyperplane, 10,000 coefficients are the values of each case that divide the discriminant coefficient space into finite CP by 100 linear hyperplanes. The interior points of each CP correspond to the discriminant coefficient that discriminates the same cases correctly and misclassifies the other case. Therefore, because the interior points of each CP have unique NM, we can define OCP with MNM.
2. Let us assume that $MNM_p$ is MNM in the $p$-dimensional space. MNM decreases monotonously ($MNM_p \geq MNM_{(p+1)}$). If $MNM_p = 0$, all MNMs, including these $p$-variables (genes), are zero. However, IP-OLDF can find the correct vertex of OCP if the data are general positions, and it might not find the correct vertex of OCP if the data are not general positions. Therefore, I develop Revised IP-OLDF that finds the OCP interior point directly. There are four serious problems in discriminant analysis before 2014. I regret having spent much research time solving these problems by Revised IP-OLDF. After establishing the Theory, I discriminate six microarray datasets by seven LDFs, with the exception of logistic regression that cannot discriminate the microarray dataset now. Because NMs of Fisher's LDF are not zero, it is not used for the datasets. Although NMs of H-SVM are zero, all coefficients of H-SVM are not zero. Therefore, H-SVM is not helpful for feature-selection. Because several coefficients of Revised IP-OLDF are not zero, and most of the coefficients are zero, Revised IP-OLDF can select features of the dataset within a few seconds. Many researchers have struggled to analyze the dataset by common statistical methods because there are many variables (genes). Recently, some researchers expect LASSO is helpful for feature-selection of the datasets. However, I find that dataset consists of several disjoint unions of SMs, with MNMs of zero. I call the linearly separable models, "Matroska" in gene analysis. We can analyze these

SMs easily because each Matroska is a small sample. However, because LSD discrimination is no longer popular, I explain Method 2 by Swiss banknote data in Chap. 6 and Japanese-automobile data in Chap. 7 before introducing Method 2 in Chap. 8.

## 1.9 Summary

In this chapter, we explained the new Theory that can discriminate several types of real data exactly. Revised IP-OLDF can solve Problems 1, 2, and 5. Problem 3 is the defect of the generalized inverse matrix technique based on the variance–covariance matrices. It is only concerned with QDF now. We can find the reason for Problem 3 with a $t$ test after three years of investigation and can solve the problem by adding slight noise to constant variables. Although discriminant analysis is not inferential statistics (Problem 4), the Theory 1 offers a simple and powerful model selection procedure, such as the best model. The best model of Revised IP-OLDF is almost the minimum $M2$ among eight LDFs. Moreover, we obtain the 95 % CI of the error rate and discriminant coefficient. Seven LDFs, with the exception of Fisher's LDF, become trivial LDF. Only Revised IP-OLDF and H-SVM can discriminate LSD theoretically. However, because H-SVM cannot discriminate data with "MNM $\geq 1$," we believe that there is no research on LSD discrimination. We investigate four types of LSD, such as Swiss banknote data, pass/fail determination of examination scores, student linearly separable data, and Japanese-automobile data. When we discriminate six microarray datasets, we find that only Revised IP-OLDF can select features naturally, and the dataset has the Matroska structure. By this fact, we can analyze high-dimensional dataset with common statistical methods such as $t$ test, one-way ANOVA, cluster analysis, and PCA easily. We hope that the Theory, Method 1, and Method 2 will be helpful for gene analysis. We confirm our theory and methods with different types of datasets from Chaps. 2–8 and prove that our theory and methods can solve many problems and find new facts of discriminant analysis.

Chapter 2: Iris Data and Fisher's Assumption

Fisher proposed Fisher's LDF under Fisher's assumption and evaluated Fisher's LDF by these data. In this book, our main policy of discrimination consists of two parts as follows:

1. We discriminate the original data by six MP-based LDFs and four statistical discriminant functions including QDF and RDA.
2. We generate resampling samples from the original data and discriminate the resampling samples by Method 1. We compare eight LDFs by $M2$ and the 95 % CI of the coefficients. We explain the LINGO Program 2 of six LDFs in Chap. 9. Because there is a small difference among the seven NMs except for H-SVM, we should no longer use Iris data for the evaluation of discriminant functions. If the data satisfy Fisher's assumption, NM of Fisher's LDF continues to converge

on MNM. Although there is no actual test for Fisher's assumption, we can confirm Fisher's assumption by this idea.

Chapter 3: CPD Data with Collinearities

In this chapter, we discriminate CPD data with three collinearities. We explain how to find collinearities and remove the effect of such collinearities. We show the strange trend of NMs by QDF and find that QDF is fragile for collinearities. Moreover, NM of Fisher's LDF does not decrease in the 19 models from the one to 19-variable model selected by the forward stepwise procedure. On the other hand, NMs of our three MP-based LDFs decrease. In the original CPD data, we select the four-variable model as useful. However, the best model recommends the nine-variable model. We believe that many variables and/or collinearities cause this difference. Because CPD data have many OCPs, Revised IP-OLDF might search for several OCPs with the same MNMs and different coefficient groups that belong to different OCPs. This result means that it is difficult for us to evaluate the 95 % CI of discriminant coefficients that might be the new Problem 6 in future work.

Chapter 4: Student data and Problem 1

The Student data consist of 40 students with six variables, which makes the data small. Although we never believe that these data might be helpful for our research, we find the defect of IP-OLDF (Problem 1). Therefore, we develop Revised IP-OLDF. Moreover, we can demonstrate that seven LDFs are quite different using a scatter plot in Fig. 4.1.

Chapter 5: Pass/Fail Determination using Examination Scores—A Trivial LDF

These data are LSD, and there is trivial LDF. In this chapter, we set the intercept to one for seven LDFs and obtain several good results, as follows:

1. $M2$ of Fisher's LDF is over 4.6 % worse than Revised IP-OLDF.
2. SVM1 is worse than another MP-based LDFs and logistic regression.
3. The 95 % CI of the best discriminant coefficients is obtained.
4. Furthermore, if we select the median of the coefficient of seven LDFs, with the exception of Fisher's LDF, seven medians are almost the same as the trivial LDF for linearly separable models.

Chapter 6: Best model of Swiss Banknote Data—Explanation 1 of Matroska Feature-selection Method (Method 2)

Swiss banknote data are LSD. We find that the two-variable model, such as ($X4$, $X6$), is the minimum model that is linearly separable; we also find that 16 models, including these two variables, are linearly separable, whereas 47 other models are not linearly separable. Therefore, we compare eight LDFs by the best models and obtain good results. Although we have not been able to explain the useful meaning of the 95 % CI of the coefficient to this point before 2014, the pass/fail determination using examination scores provides a clear understanding by normalizing the coefficient. Therefore, we attempt to explain the meaning of these data (Shinmura 2015a). Moreover, we study LSD discrimination by these data, the Japanese-automobile data, and 18 pass/fail determinations precisely. We propose

## 1.9 Summary

Method 2. Because LSD discrimination is no longer popular, we explain the Method 2 by the detailed examples of these and the Japanese-automobile data.

Chapter 7: Japanese-automobile Data—Explanation 2 of Matroska Feature-selection (Method 2)

In this chapter, we discriminate the Japanese-automobile data that are LSD. These data are a good example of Problem 3. Moreover, we can explain Method 2. Although BGSs can explain the structure of Matroska completely, two BGSs such as the capacity and emission can explain the structure of these data. Therefore, we can understand Method 2 by these data.

Chapter 8: Matroska Feature-selection Method for Microarray Data (Method 2)

In this chapter, we propose Method 2 for the datasets. We already established the Theory and developed Revised IP-OLDF. This LDF can select gene features naturally. It finds that the datasets consist of disjoint unions of several SMs. Therefore, we do not need to struggle with the high-dimensional gene space (Problem 5). If we can develop Revised LINGO Program 3 of Method 2 that can find all BGSs, it will be more useful in gene analysis. LINGO Program 3 is useful for other gene dataset, such as RNA-Seq., in addition to the microarray datasets. Because we were successful to prove the effectiveness of Maruyama vaccine (SSM) administration (Shinmura et al. 1987, Shinmura 2001), "Ancer 20 that is one of SSM" have been approved as a formal "Pharmaceuticals" since August 1991 (Noda et al. 2006). However, our survey failed to clarify the long-term survivors of SSM administration patients. If we compare two lists of cancer genes (normal and cancer patient data) versus (normal and SSM Administration patient data), and find the differences between two gene lists, it may show the proof of the effectiveness of SSM. This approach will be helpful for the effect judgment of other cancer treatment except for the surgery. We hope the advice or provision of dataset by medical doctors or project. We would like to propose a joint research in the world.

Chapter 9: This chapter explains LINGO Program 2 of Method 1.

## References

Alon U, Barkai N, Notterman DA, Gish K, Ybarra S, Mack D, Levine AJ (1999) Broad patterns of gene expression revealed by clustering analysis of tumor and normal colon tissues probed by oligonucleotide arrays. Proc Natl Acad Sci USA 96(12):6745–6750

Anderson E (1945) The irises of the Gaspe Peninsula. Bull Am Iris Soc 59:2–5

Buhlmann P, Geer AB (2011) Statistics for high-dimensional data-method, theory and applications. Springer, Berlin

Chiaretti S, Li X, Gentleman R, Vitale A, Vignetti M, Mandelli F, Ritz J, Foa R (2004) Gene expression profile of adult T-cell acute lymphocytic leukemia identifies distinct subsets of patients with Different response to therapy and survival. Blood 103/7: 2771–2778. 1 April 2004

Cox DR (1958) The regression analysis of binary sequences (with discussion). J Roy Stat Soc B 20:215–242

Firth D (1993) Bias reduction of maximum likelihood estimates. Biometrika 80:27–39

Fisher RA (1936) The Use of multiple measurements in taxonomic problems. Annals Eugenics 7:179–188

Fisher RA (1956) Statistical methods and statistical inference. Hafner Publishing Co, New Zealand

Flury B, Rieduyl H (1988) Multivariate statistics: a practical approach. Cambridge University Press, New York

Friedman JH (1989) Regularized discriminant analysis. J Am Stat Assoc 84(405):165–175

Glover L (1990) Improvement linear programming models for discriminant analysis. Decis Sci 2:771–785

Golub TR, Slonim DK, Tamayo P, Huard C, Gaasenbeek M, Mesirov JP, Coller H, Loh ML, Downing JR, Caligiuri MA, Bloomfield CD, Lander ES (1999) Molecular classification of cancer: class discovery and class prediction by gene expression monitoring. Science 286 (5439):531–537

Goodnight JH (1978) SAS technical report—the sweep operator: its importance in statistical computing—(R100). SAS Institute Inc, USA

Ishii A, Yata K, Aoshima M (2014) Asymptotic distribution of the largest eigenvalue via geometric representations of high-dimension, low-sample-size data. Sri Lankan J Appl Statist, Special issue: modern statistical methodologies in the cutting edge of science (ed. Mukhopadhyay, N.): 81–94

Ibaraki T, Muroga S (1970) Adaptive linear classifier by linear programming. IEEE Trans Syst Sci Cybern 6(1):53–62

Jeffery IB, Higgins DG, Culhane C (2006) Comparison and evaluation of methods for generating differentially expressed gene lists from microarray data. BMC Bioinf 7:359: 1–16. doi:10.1186/1471-2105-7-359

Konishi S, Honda M (1992) Bootstrap methods for error rate estimation in discriminant analysis. Jpn Soc Appl Stat 21(2):67–100

Lachenbruch PA, Mickey MR (1968) Estimation of error rates in discriminant analysis. Technometrics 10:1–11

Liitschwager JM, Wang C (1978) Integer programming solution of a classification problem. Manage Sci 24(14):1515–1525

Markowitz HM (1959) Portfolio selection, efficient diversification of investment. Wiley, USA

Miyake A, Shinmura S (1976) Error rate of linear discriminant function. In: Dombal FT, Gremy F (ed). North-Holland Publishing Company, The Netherland, pp 435–445

Miyake A, Shinmura S (1979) An algorithm for the optimal linear discriminant functions. Proceedings of the international conference on cybernetics and society, pp 1447–1450

Miyake A, Shinmura S (1980) An algorithm for the optimal linear discriminant function and its application. Jpn Soc Med Electron Biol Eng 18(6):452–454

Noda K, Ohashi Y, Okada H, Ogita S, Ozaki M, Kikuchi Y, Takegawa Y, Niibe H, Fujii S, Horiuchi J, Morita K, Hashimoto S, Fujiwara K (2006) Randomized phase II study of immunomodulator Z-100 in patients with stage IIIB cervical cancer with radiation therapy. Jpn J Clin Oncol 36(9):570–577 Epub 2006 Aug 22

Nomura Y, Shinmura S (1978) Computer-assisted prognosis of acute myocardial infarction. MEDINFO 77, In: Shires W (ed) IFIP. North-Holland Publishing Company, The Netherland, pp 517–521

Rubin PA (1997) Solving mixed integer classification problems by decomposition. Ann Oper Res 74:51–64

Sall JP (1981) SAS regression applications. SAS Institute Inc., USA. (Shinmura S. translate Japanese version)

Sall JP, Creighton L, Lehman A (2004) JMP start statistics, third edition. SAS Institute Inc., USA. (Shinmura S. edits Japanese version)

Schrage L (1991) LINDO—an optimization modeling systems. The Scientific Press, UK. (Shinmura S. & Takamori, H. translate Japanese version)

Schrage L (2006) Optimization modeling with LINGO. LINDO Systems Inc., USA. (Shinmura S. translates Japanese version)

# References

Shimizu T, Tsunetoshi Y, Kono H, Shinmura S (1975) Classification of subjective symptoms of junior high school Students affected by photochemical air pollution. J Jpn Soc Atmos Environ 9/4: 734–741. Translated for NERC Library, EPA, from the original Japanese by LEO CANCER Associates, P.O.Box 5187 Redwood City, California 94063, 1975 (TR 76-213)

Shinmura S, Kitagawa M, Takagi Y, Nomura Y (1973) The spectrum diagnosis by a two-stage weighting (nidankain omomizukeniyoru supekutoru sindan). The 12th conference of BME: 107–108

Shinmura S, Kitagawa M, Nomura Y (1974) The spectrum diagnosis (Part 2). The 13th conference of BME, pp 414–415

Shinmura S, Miyake A (1979) Optimal linear discriminant functions and their application. COMPSAC 79:167–172

Shinmura S, Suzuki T, Koyama H, Nakanishi K (1983) Standardization of medical data analysis using various discriminant methods on a theme of breast diseases. MEDINFO 83, In: Vann Bemmel JH, Ball MJ, Wigertz O (ed). North-Holland Publishing Company, The Netherland, pp 349–352

Shinmura S (1984) Medical data analysis, model, and OR. Oper Res 29(7):415–421

Shinmura S, Iida K, Maruyama C (1987) Estimation of the effectiveness of cancer treatment by SSM using a null hypothesis model. Inform Health Social Care 7(3):263–275. doi:10.3109/14639238709010089

Shinmura S (1998) Optimal linear discriminant functions using mathematical programming. J Jpn Soc Comput Stat 11(2):89–101

Shinmura S, Tarumi T (2000) Evaluation of the optimal linear discriminant functions using integer programming (IP-OLDF) for the normal random data. J Jpn Soc Comput Stat 12(2):107–123

Shinmura S (2000a) A new algorithm of the linear discriminant function using integer programming. New Trends Prob Stat 5:133–142

Shinmura S (2000b) Optimal linear discriminant function using mathematical programming. Dissertation, March 200: 1–101, Okayama University, Japan

Shinmura S (2001) Analysis of effect of SSM on 152,989 cancer patient. ISI2001: 1–2. doi:10.13140/RG.2.1.30779281

Shinmura S (2003) Enhanced algorithm of IP-OLDF. ISI2003 CD-ROM, pp 428–429

Shinmura S (2004) New algorithm of discriminant analysis using integer programming. IPSI 2004 Pescara VIP Conference CD-ROM, pp 1–18

Shinmura S (2005) New age of discriminant analysis by IP-OLDF—Beyond Fisher's linear discriminant function. ISI2005, pp 1–2

Shinmura S (2007) Overviews of discriminant function by mathematical programming. J Jpn Soc Comput Stat 20(1–2):59–94

Shinmura S (2010a) The optimal linearly discriminant function (Saiteki Senkei Hanbetu Kansuu). Union of Japanese Scientist and Engineer Publishing, Japan

Shinmura S (2010b) Improvement of CPU time of Revised IP-OLDF using linear programming. J Jpn Soc Comput Stat 22(1):39–57

Shinmura S (2011a) Beyond Fisher's linear discriminant analysis—new world of the discriminant analysis. ISI2011 CD-ROM, pp 1–6

Shinmura S (2011b) Problems of discriminant analysis by mark sense test data. Jpn Soc Appl Stat 40(3):157–172

Shinmura S (2013) Evaluation of optimal linear discriminant function by 100-fold cross-validation. ISI2013 CD-ROM, pp 1–6

Shinmura S (2014a) End of discriminant functions based on variance-covariance matrices. ICORE2014, pp 5–16

Shinmura S (2014b) Improvement of CPU time of linear discriminant functions based on MNM criterion by IP. Stat, Optim Inf Comput 2:114–129

Shinmura S (2014c) Comparison of linear discriminant functions by $K$-fold cross-validation. Data Anal 2014:1–6

Shinmura S (2015a) The 95 % confidence intervals of error rates and discriminant coefficients. Stat, Optimi Inf Comput 2:66–78

Shinmura S (2015b) A trivial linear discriminant function. Stat, Optim Inf Comput 3:322–335. doi:10.19139/soic.20151202

Shinmura S (2015c) Four serious problems and new facts of the discriminant analysis. In: Pinson E, Valente F, Vitoriano B (ed) Operations research and enterprise systems, pp 15–30. Springer, Berlin (ISSN: 1865-0929, ISBN: 978-3-319-17508-9. doi:10.1007/978-3-319-17509-6)

Shinmura S (2015d) Four Problems of the Discriminant Analysis. ISI2015:1–6

Shinmura S (2015e) The Discrimination of microarray data (Ver. 1). Res Gate (1) 1–4. 28 Oct 2015

Shinmura S (2015f) Feature-selection of three microarray data. Research Gate (2) 1–7. 1 Nov 2015

Shinmura S (2015g) Feature-selection of microarray data (3)—Shipp et al. microarray data. Res Gate 3:1–11

Shinmura S (2015h) Validation of feature-selection (4)—Alon et al. microarray data. Res Gate 4:1–11

Shinmura S (2015i) Repeated feature-selection method for microarray data (5). Res Gate (5) 1–12. 9 Nov 2015

Shinmura S (2015j) Comparison Fisher's LDF by JMP and Revised IP-OLDF by LINGO for microarray data (6). Res Gate (6)1–10. 11 Nov 2015

Shinmura S (2015k) Matroska trap of feature-selection method (7)—Golub et al. microarray data. Res Gate (7) 1–14. 18 Nov 2015

Shinmura S (2015l) Minimum sets of genes of Golub et al. microarray data (8). Res Gate (8) 1–12. 22 Nov 2015

Shinmura S (2015m) Complete lists of small matroska in Shipp et al. microarray data (9). Res Gate (9) 1–81. 4 Dec 2015

Shinmura S (2015n) Sixty-nine small matroska in Golub et al. microarray data (10). Res Gate 1–58

Shinmura S (2015o) Simple structure of Alon et al. et al. microarray data (11). Res Gate(11) 1–34. 4 Dec 2015

Shinmura S (2015p) Feature-selection of Singh et al. microarray data (12). Res Gate (12) 1–89. 6 Dec 2015

Shinmura S (2015q) Final list of small matroska in Tian et al. microarray data. Res Gate (13) 1–160

Shinmura S (2015r) Final list of small matroska in Chiaretti et al. microarray data. Res Gate (14) 1–16. 20 Dec 2015

Shinmura S (2015s) Matroska feature-selection method for microarray data. Res Gate (15) 1–16. 20 Dec 2015

Shinmura S (2016a) The best model of Swiss banknote data. Stat Optim Inf Comput, 4: 118–131. International Academic Press (ISSN: 2310-5070 (online) ISSN: 2311-004X (print), doi:10.19139/soic.v4i2.178)

Shinmura S (2016b) Matroska feature-selection method for microarray data. Biotechnology 2016:1–8

Shinmura S (2016c) Discriminant analysis of the linear separable data—Japanese-automobiles. J Stat Sci Appl X, X: 0–14

Shipp MA, Ross KN, Tamayo P, Weng AP, Kutok JL, Aguiar RC, Gaasenbeek M, Angelo M, Reich M, Pinkus GS, Ray TS, Koval MA, Last KW, Norton A, Lister TA, Mesirov J, Neuberg DS, Lander ES, Aster JC, Golub TR (2002) Diffuse large B-cell lymphoma outcome prediction by gene-expression profiling and supervised machine learning. Nat Med 8(1):68–74. doi:10.1038/nm0102-68

Simon N, Friedman J, Hastie T, Tibshirani R (2013) A sparse-group lasso. J Comput Graph Statist 22:231–245

Singh D, Febbo PG, Ross K, Jackson DG, Manola J, Ladd C, Tamayo P, Renshaw AA, D'Amico AV, Richie JP, Lander ES, Lada M, Kantoff PW, Golub TR, Sellers WR (2002) Gene expression correlates of clinical prostate cancer behavior. Cancer Cell 1/2: 203–209

Stam A (1997) Non-traditional approaches to statistical classification: Some perspectives on Lp-norm methods. Ann Oper Res 74:1–36

# References

Taguchi G, Jugular R (2002) The Mahalanobis-Taguchi Strategy—a pattern technology system. Wiley, New York

Tian E, Zhan F, Walker R, Rasmussen E, Ma Y, Barlogie B, Shaughnessy JD (2003) The role of the Wnt-signaling Antagonist DKK1 in the development of osteolytic lesions in multiple myeloma. New England J Med 349(26):2483–2494

VapnikV (1995) The nature of statistical learning theory. Springer, Berlin

Warmack RE, Gonzalez RC (1973) An algorithm for the optimal solution of linear inequalities and its application to pattern recognition. IEEE Transac Comput C-2(12):1065–1075

# Chapter 2
# Iris Data and Fisher's Assumption

## 2.1 Introduction

Anderson (1945) collected Iris data with three species, setosa, virginica, and vercicolor that have four (independent) variables. Because Fisher (1936, 1956) evaluated Fisher's linear discriminant function (Fisher's LDF) with these data, such data have been very popular for the evaluation of discriminant functions. Therefore, we call these data, "Fisher's Iris data." Because we can easily separate setosa from virginica and vercicolor through a scatter plot, we usually discriminate two classes, such as the virginica and vercicolor. In this book, our main policy of discrimination consists two parts: (1) evaluation of Iris data by five MP-based LDFs, with the exception of hard-margin SVM (H-SVM) by LINGO Program 1 (Schrage 2006), and four statistical discriminant functions such as Fisher's LDF, logistic regression (Cox 1958), QDF, and RDA (Friedman 1989) by JMP (Sall et al. 2004); (2) comparison of seven LDFs by the 100-fold cross-validation for small sample method (Method 1) by LINGO Program 2 and JMP script.

### 2.1.1 Evaluation of Iris Data

First, we discriminate Iris data by five MP-based LDFs, with the exception of H-SVM, and four statistical discriminant functions. LINGO solves six MP-based LDFs, such as Revised IP-OLDF based on MNM criterion (Miyake and Shinmura 1979, 1980, Shinmura 2010a, 2011a, b), Revised LP-OLDF, Revised IPLP-OLDF (Shinmura 2010b, 2014b), SVM4 (penalty $c$ = 10,000), and SVM1 (penalty $c$ = 1) (Vapniik 1995). Section 2.3.3 describes a LINGO Program 1 of six MP-based LDFs for conventional data. Downloading a free version of LINGO with manual from

LINDO Systems Inc.[1] allows anyone to analyze small sample. JMP discriminated the data by QDF and RDA (Friedman 1989), in addition to two LDFs, such as Fisher's LDF and logistic regression. Because the JMP division in Japan offers a free trial version, the reader must confirm to download JMP trial version in each country. The JMP sample folder includes Iris data. I evaluate these nine discriminant functions by NM and discuss the difference of the discriminant coefficient. Because the discriminant analysis is not the inferential statistics (Problem 4) (Shinmura 2014a, 2015c, d), we discuss model selection by regression analysis. If we use "$y_i$" [le: an indicator of MP-based LDF in Eq. (1.8)] as the object variable of the regression analysis, the obtained regression coefficients are proportional to the coefficient of Fisher's LDF (plug-in rule1). Therefore, we can use a stepwise procedure and statistics, such as AIC, BIC, and Cp statistics. We determine a good model in Iris data. However, these statistics select different good models by other data. The result is a limitation in the evaluation using a small sample (i.e., training sample). Section 2.3.3 describes a LINGO model of six MP-based LDFs.

### 2.1.2 100-Fold Cross-Validation for Small Sample (Method 1)

We generate resampling samples from Iris data and discriminate these samples by the Method 1 (Shinmura 2010a, 2013, 2014c). There is an explanation of the LINGO Program 2 of the Method 1 in Chap. 9. Because we omit QDF and RDA, we compare five MP-based LDFs and Fisher's LDF and logistic regression by two means of error rates, M1 and M2, in the training and validation samples, respectively, and the 95 % confidence interval (CI) of error rate and discriminant coefficients.

We compare seven LDFs by M2 and 95 % CI of discriminant coefficients. Because M1 of Revised IP-OLDF is the average of 100 MNMs in the training samples, it decreases monotonously similarly to MNM, and MNM of the full model is always minimum. Therefore, we cannot use M1 of Revised IP-OLDF for the model (feature or variable) selection procedure. We propose the direct and powerful model selection procedure performed by the best model with the minimum M2 in each LDF (Shinmura 2016a, b). We should select the best model by this simple model selection procedure instead of LOO procedure (Lachenbruch and Mickey 1968). We examined the best model of Revised IP-OLDF with the minimum M2 among all LDFs. There is a small difference among the seven best models because Iris data can satisfy Fisher's assumption. Although we explain Program 2 in Chap. 9, we omit the JMP script of the Method 1 for Fisher's LDF and logistic regression because explanation of the JMP script language is outside the scope of this book. Many researchers use Iris data to evaluate the discriminant function. However, we should not use these data for assessment because there is a small difference among

---

[1]http://www.lindo.com/.

## 2.1 Introduction

discriminant functions. Fisher proposed Fisher's LDF under Fisher's assumption. However, there are no actual test statistics to determine whether the data satisfy Fisher's assumption. If data satisfy Fisher's assumption, NM of Fisher's LDF continues to converge on MNM. We can confirm this fact by this idea. However, these data are critical for correlation education because we erroneously judge the meaning of correlation without verification from a scatter plot. Moreover, Anscombe's quartet is critical for correlation and simple regression education, also.

## 2.2 Iris Data

### 2.2.1 Data Outlook

Iris data have been critical evaluation data for discriminant analysis until now. Such data consist of three species: setosa, versicolor, and virginica. Each species has 50 cases, as listed in Table 2.1. There are four (independent) variables: $X1$ (petal width), $X2$ (petal length), $X3$ (sepal width), and $X4$ (sepal length). The lengths are in millimeters. Left four columns are setosa, middle four columns are versicolor, and right four columns are virginica. We can find and download from the Research Gate.[2]

Figure 2.1 shows a matrix of scatter plots of three species. A total of 50 setosa are shown in red color and the symbol ".". A total of 50 virginica are shown in green and the symbol "×". A total of 50 vercicolor are shown in blue and the symbol "+". The scatter plot shown in the first row and fourth column is $X1$ (Sepal length) by $X4$ (Petal width). Setosa, virginica, and vercicolor are located on the left, middle, and right area, respectively. We can easily separate setosa from virginica and vercicolor because two sepal lengths and petal widths of setosa are smaller than the others. Therefore, we can quickly separate setosa from the other two species through the scatter plot. We usually omit setosa and focus on the discrimination of

**Table 2.1** Iris data

| SN | X1 | X2 | X3 | X4 | X1 | X2 | X3 | X4 | X1 | X2 | X3 | X4 |
|---|---|---|---|---|---|---|---|---|---|---|---|---|
| 1 | 5.1 | 3.5 | 1.4 | 0.2 | 7 | 3.2 | 4.7 | 1.4 | 6.3 | 3.3 | 6 | 2.5 |
| 2 | 4.9 | 3 | 1.4 | 0.2 | 6.4 | 3.2 | 4.5 | 1.5 | 5.8 | 2.7 | 5.1 | 1.9 |
| ⋮ | ⋮ | ⋮ | ⋮ | ⋮ | ⋮ | ⋮ | ⋮ | ⋮ | ⋮ | ⋮ | ⋮ | ⋮ |
| 49 | 5.3 | 3.7 | 1.5 | 0.2 | 5.1 | 2.5 | 3 | 1.1 | 6.2 | 3.4 | 5.4 | 2.3 |
| 50 | 5 | 3.3 | 1.4 | 0.2 | 5.7 | 2.8 | 4.1 | 1.3 | 5.9 | 3 | 5.1 | 1.8 |
| MIN | 4.3 | 2.3 | 1 | 0.1 | 4.9 | 2 | 3 | 1 | 4.9 | 2.2 | 4.5 | 1.4 |
| MEAN | 5.01 | 3.43 | 1.46 | 0.25 | 5.94 | 2.77 | 4.26 | 1.33 | 6.59 | 2.97 | 5.55 | 2.03 |
| MAX | 5.8 | 4.4 | 1.9 | 0.6 | 7 | 3.4 | 5.1 | 1.8 | 7.9 | 3.8 | 6.9 | 2.5 |
| STD | 0.35 | 0.38 | 0.17 | 0.11 | 0.52 | 0.31 | 0.47 | 0.2 | 0.64 | 0.32 | 0.55 | 0.27 |

---

[2] https://www.researchgate.net/profile/Shuichi_Shinmura.

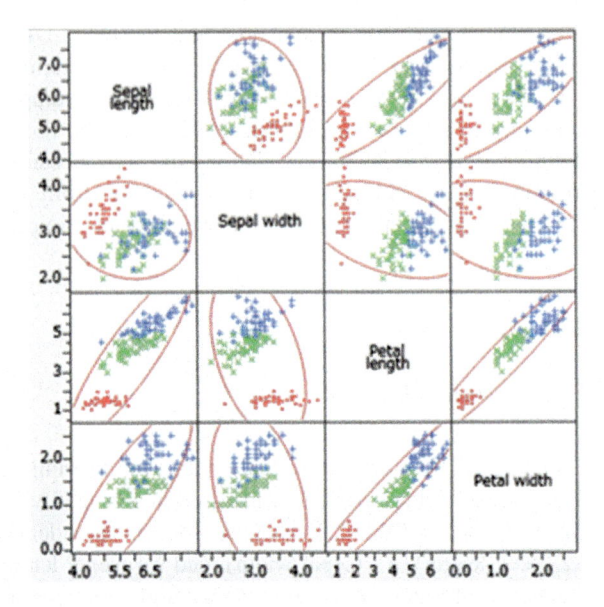

**Fig. 2.1** Matrix of scatter plots of Iris data (three species)

versicolor and virginica. Iris data are linearly separable data (LSD) between setosa and others. Because we can find Iris data are LSD easily, we never discuss the importance and difficulties of LSD discrimination. We might erroneously assume that LSD discrimination can be performed easily by the impression of Iris data.

Figure 2.2 is a matrix scatter plot of two species. A total of 50 virginica are shown in green color and with the symbol "×". A total of 50 vercicolor are shown in blue and with the symbol "+". Virginica and vercicolor are located on the left and

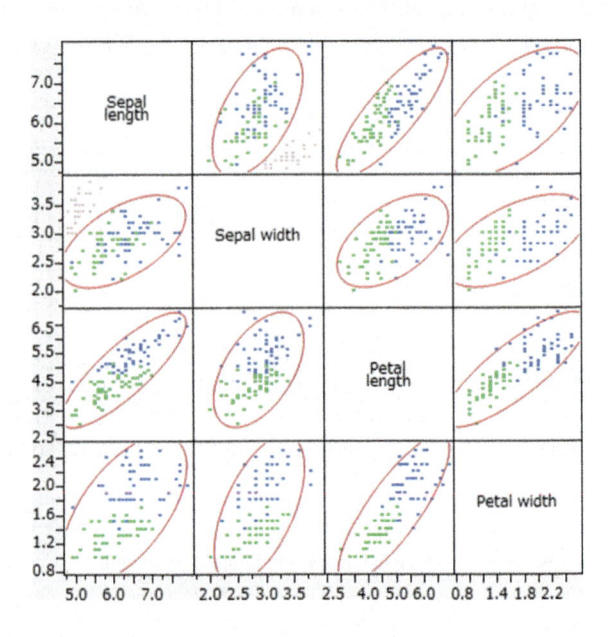

**Fig. 2.2** Scatter plots of Iris data (two species)

## 2.2 Iris Data

right area, respectively. Although we cannot determine whether these data satisfy Fisher's assumption, we can estimate that these data satisfy Fisher's assumption because NMs of nine discriminant functions are almost the same, as indicated in Table 2.4. Although Iris data with three species are LSD, Iris data with two species are not LSD.

Table 2.2 lists two correlation matrices that correspond to Figs. 2.1 (left) and 2.2 (right). Although the correlation of ($X2$, $X4$) in the three species is −0.37, the correlation of ($X2$, $X4$) in the two species is 0.57. These data are good educational examples of correlation because we erroneously judge correlation without the verification of the scatter plot. In Fig. 2.1, because the scatter plot indicates that setosa is completely apart from the other species, we never consider the correlation shown in Fig. 2.1 and indicated in Table 2.2 (left). On the other hand, all scatter plots include most cases of two species in the 95 % probability ellipse. In these cases, the correlations are meaningful.

### 2.2.2 Model Selection by Regression Analysis

Although the discriminant analysis is not the inferential statistics, there is no correct model selection procedure aside from LOO procedure. We can use the stepwise procedure of the regression analysis and several statistics, such as AIC, BIC, and Cp, by the plug-in rule1. To this point, we have two options for selecting a good model for the original data. The first is the LOO procedure. The second is to evaluate the models by the model selection statistics of regression analysis. Table 2.3 lists the result of all possible model combinations (Goodnight 1978). The column "Model" lists 15 models from four- and one-variable models. "$p$" indicates the number of variables. Within the same "$p$", the models are in descending order of "$R$-square ($R^2$)." "Rank" is the ranking within the same number of "$p$". This procedure is very powerful because we can overlook all models and simulate the forward and backward stepwise procedures. Both procedures select the same models, such as ($X4$) → ($X4$, $X2$) → ($X4$, $X2$, $X3$) → ($X4$, $X2$, $X3$, $X1$). Therefore, we can easily select a good model among these 15 models. Model selection statistics, such as AIC, BIC, and Cp, select the full model as a good model. However, these statistics usually select different models for other data. Therefore, we cannot usually determine a good model from these statistics alone.

**Table 2.2** Two correlation matrices

|  | Three species in Fig. 2.1 ||||  Two species in Fig. 2.2 ||||
|---|---|---|---|---|---|---|---|---|
|  | X1 | X2 | X3 | X4 | X1 | X2 | X3 | X4 |
| X1 | 1 | −0.12 | 0.87 | 0.82 | 1 | 0.55 | 0.83 | 0.59 |
| X2 | −0.12 | 1 | −0.43 | ***−0.37*** | 0.55 | 1 | 0.52 | ***0.57*** |
| X3 | 0.87 | −0.43 | 1 | 0.96 | 0.83 | 0.52 | 1 | 0.82 |
| X4 | 0.82 | ***−0.37*** | 0.96 | 1 | 0.59 | ***0.57*** | 0.82 | 1 |

**Table 2.3** Result of all possible combinations

| Model | p | Rank | $R^2$ | AIC | BIC | Cp |
|---|---|---|---|---|---|---|
| 1, 2, 3, 4 | 4 | 1 | 0.78 | *143.5* | *158.2* | 5 |
| 2, 3, 4 | 3 | 1 | 0.77 | 148.7 | 161.1 | 10.37 |
| 1, 3, 4 | 3 | 2 | 0.76 | 151.8 | 164.2 | 13.59 |
| 1, 2, 4 | 3 | 3 | 0.73 | 163.9 | 176.3 | 27.16 |
| 1, 2, 3 | 3 | 4 | 0.7 | 174.2 | 186.6 | 40.09 |
| 2, 4 | 2 | 1 | 0.72 | 163.5 | 173.5 | 27.39 |
| 3, 4 | 2 | 2 | 0.72 | 165 | 175 | 29.19 |
| 1, 3 | 2 | 3 | 0.7 | 172.7 | 182.7 | 39.07 |
| 1, 4 | 2 | 4 | 0.69 | 176.4 | 186.4 | 44.12 |
| 2, 3 | 2 | 5 | 0.63 | 192.1 | 202.1 | 67.61 |
| 1, 2 | 2 | 6 | 0.25 | 264 | 274 | 237.4 |
| 4 | 1 | 1 | 0.69 | 174.3 | 181.8 | 42.12 |
| 3 | 1 | 2 | 0.62 | 193.7 | 201.3 | 71.72 |
| 1 | 1 | 3 | 0.24 | 262 | 269.6 | 236.2 |
| 2 | 1 | 4 | 0.09 | 280.1 | 287.6 | 2.87 |

## 2.3 Comparison of Seven LDFs

### 2.3.1 Comparison of MNM and Eight NMs

We investigate all possible combinations of the discriminant models ($15 = 2^4 - 1$) by LINGO and JMP script. Table 2.4 lists the 15 models from four to one variable, similar to Table 2.3. "SN" is the sequential number of models. "Var" denotes the suffix of variables. "RIP" is the MNMs of Revised IP-OLDF. We can confirm the monotonic decrease of MNM ($MNM_p \geq MNM_{(p+1)}$). For example, the forward stepwise procedure of the regression analysis selects the following variables in this order: $X4$, $X2$, $X3$, and $X1$. MNM of the four models decreases as follows: 6, 3, 2, and 1. We can confirm the monotonic decrease of MNM by other model sequences, such as $X1$, $X2$, $X3$, and $X4$, in this order. MNM of the four models decreases as follows: 37, 25, 2, and 1. Therefore, we cannot select the model with minimum MNM as the best model because we always select the full model. This fact tells us clearly why we should not select model by statistics in the training sample. The four discriminant functions represent the following abbreviations in the table: Revised LP-OLDF is LP, Revised IPLP-OLDF is IPLP, logistic regression is "Logistic," and Fisher's LDF is LDF. The seven columns that follow "RIP" are the difference (Diff1) defined as (NMs of seven discriminant functions − MNM). All the NMs of each model should be greater than or equal to MNM because MNM is the minimum NM in the training samples. The last row shows the number of models with a negative value of "Diff1." Revised LP-OLDF has two negative values. This means that Revised LP-OLDF is fragile for Problem 1. The column "ZERO" shows the number of cases on the discriminant hyperplane by Revised

2.3 Comparison of Seven LDFs

**Table 2.4** MNM and eight "Diff1"

| SN | Var | RIP | LP | Zero | IPLP | SVM4 | SVM1 | Logistic | LDF | QDF | RDA |
|---|---|---|---|---|---|---|---|---|---|---|---|
| 1 | 1, 2, 3, 4 | 1 | 1 | | 0 | 1 | 0 | 1 | *2* | *2* | *2* |
| 2 | 2, 3, 4 | 2 | 0 | | 0 | 0 | *2* | 0 | *2* | *2* | 1 |
| 3 | 1, 3, 4 | 2 | 0 | | 0 | 0 | 0 | 0 | 1 | 1 | *2* |
| 4 | 1, 2, 4 | 4 | *3* | | 0 | *3* | 1 | 0 | 1 | 2 | 1 |
| 5 | 1, 2, 3 | 2 | 2 | 1 | 0 | 2 | 4 | 2 | 5 | *6* | 4 |
| 6 | 2, 4 | 3 | *3* | | 0 | 1 | 1 | 0 | 0 | 2 | 2 |
| 7 | 3, 4 | 5 | 1 | | 0 | *3* | 2 | 1 | 3 | 0 | 2 |
| 8 | 1, 3 | 4 | 1 | | 0 | 1 | *3* | 0 | 2 | 2 | 2 |
| 9 | 1, 4 | 6 | 0 | | 0 | *1* | *1* | 0 | *1* | 0 | 0 |
| 10 | 2, 3 | 5 | *1* | | 0 | 0 | 0 | 0 | *1* | *1* | *1* |
| 11 | 1, 2 | 25 | 2 | | 0 | 2 | 2 | 0 | 0 | *4* | *4* |
| 12 | 4 | 6 | *0* | | *0* | *0* | *0* | *0* | *0* | *0* | *0* |
| 13 | 3 | 7 | 0 | | 0 | 0 | 0 | 0 | 1 | 0 | 0 |
| 14 | 1 | 37 | −3 | 4 | 0 | 0 | 0 | 0 | 0 | *3* | *3* |
| 15 | 2 | 27 | −2 | 10 | 0 | *5* | *5* | *0* | *5* | *5* | *5* |
| n (−) | | − | 2 | | 0 | 0 | 0 | 0 | 0 | 0 | 0 |
| n (Bold) | | | 4 | | 1 | 5 | 5 | 1 | 8 | 8 | 7 |

LP-OLDF. This fact explains that we cannot determine Problem 1 for those models with "Diff1 ≥ 0." Moreover, we cannot discuss Problem 1 for the four statistical discriminant functions because all statistical software developers do not know Problem 1 and do not support this information. Therefore, we need to re-evaluate the old results of the discriminant analysis. Although these data are expected to provide the correct results for Fisher's LDF, QDF, and RDA, these three NMs are not less than those of MP-based LDFs. The bold numbers of the "Diff1s" among each of the 15 models are the maximum values. There are 23 maximum values among Fisher's LDF, QDF, and RDA. On the other hand, there are 15 maximum values among LP, IPLP, SVM4, SVM1, and Logistic. In general, we judge Fisher's LDF, QDF, and RDA to be inferior to other LDFs, although this determination is not clear because there is no appropriate threshold.

### 2.3.2 Comparison of Seven Discriminant Coefficient

Table 2.5 lists the coefficients of full models. We divide the original coefficients of five MP-based LDFs by the intercept terms in parentheses (Shinmura 2015a, b). The four coefficients are normalized. When we analyze the data by regression analysis, the obtained regression coefficients are proportional to the coefficients of Fisher's LDF. The first row of Fisher's LDF is the original regression coefficients. The second row is the normalized coefficients divided by the intercept. The third

**Table 2.5** Coefficients of seven LDFs

|  | X1 | X2 | X3 | X4 | c |
|---|---|---|---|---|---|
| RIP | 0.06 | 0.11 | −0.25 | −0.27 | (−120.25) |
| SVM4 | 0.04 | 0.24 | −0.19 | −0.57 | (−33.6) |
| SVM1 | 0.09 | 0.14 | −0.3 | −0.3 | (−6.78) |
| LP | 0.04 | 0.24 | −0.19 | −0.57 | (−33.6) |
| IPLP | 0.06 | 0.11 | −0.25 | −0.27 | (−120.25) |
| Fisher's LDF | 0.39 | 0.62 | −0.77 | −1.37 | 1.84 |
| Normalize | 0.21 | 0.34 | −0.42 | −0.74 | 1 |
| SE | 0.14* | 0.19* | 0.16* | 0.23* | 0.53* |
| Logistic | −2.47 | −6.68 | 9.43 | 18.29 | −42.64 |
| Normalize | 0.06 | 0.16 | −0.22 | −0.43 | 1 |
| SE | 2.39 | 4.48 | 4.74 | 9.74 | 25.71 |

row is the SEs of the regression coefficient. The first and second rows of logistic regression are similar to Fisher's LDF. SE of logistic regression is achieved through a numerical calculation using the maximum likelihood method proposed by R. Fisher. We can reject the four coefficients and intercept of Fisher's LDF at the 5 % level. This indicates that the full model of Fisher's LDF is selected by the 95 % CI of coefficients. On the other hand, we cannot reject the four coefficients of the logistic regression at the 5 % level. This fact implies us the 95 % CIs of logistic coefficient are not reliable. Moreover, we cannot discuss the 95 % CI of the coefficient by MP-based LDFs in the original data. If we focus on the normalized coefficient values, only values of Fisher's LDF are greater than the other coefficients. We suggest that statistical software developers support the Method 1 and an option for setting the intercept to one. This is helpful to users because users cannot obtain useful information through discriminant analysis.

### 2.3.3 LINGO Program 1: Six MP-Based LDFs for Original Data

We introduce the LINGO model for six MP-based LDFs, including H-SVM. This Program 1 can be used for a small change in the underlined parts that consist of four critical sections. The "SETS" section defines one-dimensional sets and arrays by "set: arrays." Because the Iris data have four variables, set "P" has four elements. Set "P1" has five elements, including the intercept; the fifth element stores intercept "$y_i$," and "VARK" is the five-element array that corresponds to set "P1." Set "P2" has six elements; the sixth element stores the sequential model number from one to 15. Set "N" has 100 elements of the number of cases. Array "E" stores 0/1 decision variable "$e_i$", and array "CONSTANT" stores the values of the discriminant scores. Set "G2"

## 2.3 Comparison of Seven LDFs

has five elements that correspond to the five MP-based LDFs, with the exception of H-SVM. Set "MS" has 15 elements that correspond to all possible models. Set "V2" has 75 elements (=15 models × 5 LDFs) that store 75 coefficients. Set "MB" made by two one-dimensional sets is a two-dimensional set with 15 rows by five columns with intercept. Array "CHOICE" defines the variable in 15 models by 0/1 values, similar to what is shown in Fig. 9.4. Set "ERR" is a two-dimensional set with 15 rows by five columns. Array "IC" stores NM and the "ZERO" that stores the number of cases on the discriminant hyperplane indicated in Table 2.4. Set "VP" is a two-dimensional set with 75 rows by six columns, and array "VARK75" stores the coefficients. Set "D" is a two-dimensional set with 100 rows by five columns, and array "IS" defines the modified Iris data indicated in Table 2.6 (right). JMP uses the left five columns. LINGO uses the right five columns (the cell range name "IS" is defined by "G2: K101" in Excel); the values of class2 have negative values.

```
MODEL:
SETS:
  P; P1:VARK; P2;
  N: E, CONSTANT; G2:; MS:; V2:;
  MB(MS, P1): CHOICE;
  ERR(MS, G2): IC, ZERO;
  VP(V2, P2): VARK75;
  D(N, P1): IS;
ENDSETS
```

The DATA section defines the number of elements for "P, P1, P2, N, G2, MS, V2, G24," and the constant values, such as "NN and BIGM." "P=1..4;" means set "P" has four elements. If we change the underlined parts of the data, the LINGO model can discriminate such data. The most important role is to input the data from

**Table 2.6** Iris data for JMP (*left*) and LINGO (*right*)

| SN | X1 | X2 | X3 | X4 | Species | X1 | X2 | X3 | X4 | $y_i$ |
|---|---|---|---|---|---|---|---|---|---|---|
| 1 | 7 | 3.2 | 4.7 | 1.4 | Versicolor | 7 | 3.2 | 4.7 | 1.4 | 1 |
| 2 | 6.4 | 3.2 | 4.5 | 1.5 | Versicolor | 6.4 | 3.2 | 4.5 | 1.5 | 1 |
| : | : | : | : | : | : | : | : | : | : | : |
| 49 | 5.1 | 2.5 | 3 | 1.1 | Versicolor | 5.1 | 2.5 | 3 | 1.1 | 1 |
| 50 | 5.7 | 2.8 | 4.1 | 1.3 | Versicolor | 5.7 | 2.8 | 4.1 | 1.3 | 1 |
| 51 | 6.3 | 3.3 | 6 | 2.5 | Virginica | −6.3 | −3.3 | −6 | −2.5 | −1 |
| 52 | 5.8 | 2.7 | 5.1 | 1.9 | Virginica | −5.8 | −2.7 | −5.1 | −1.9 | −1 |
| : | : | : | : | : | : | : | : | : | : | : |
| 99 | 6.2 | 3.4 | 5.4 | 2.3 | Virginica | −6.2 | −3.4 | −5.4 | −2.3 | −1 |
| 100 | 5.9 | 3 | 5.1 | 1.8 | Virginica | −5.9 | −3 | −5.1 | −1.8 | −1 |

a database, such as Excel, and array name, such as "IS = @OLE();" and output the results for the same database, such as Excel, by "@OLE() = IC, ZERO, NP, VARK75;" LINGO and Excel share the same array name.

```
DATA:
  P=1..4; P1=1..5; P2=1..6;
  N=1..100; G2=1..5; MS=1..15; V2=1..75;
  G24=1..1500; NN=100; BIGM=10,000 (or 1);
  CHOICE, IS=@OLE();
ENDDATA
```

The SUBMODEL section defines five MP-based LDFs, such as Revised IP-OLDF represented by the submodel name "RIP," and soft-margin support vector machine (S-SVM) represented by the submodel name "SVM." If we set "BIGM = 10,000," submodel "SVM" becomes SVM4 in the CALC section.

```
SUBMODEL RIP:
 MIN=ER;
 ER=@SUM(N(i):E(i));
 @FOR(N(i):@SUM(P1(J1):IS(i,J1)*VARK(J1)*CHOICE(k,J1)) > 1-BIGM*E(i));
 @FOR(P1(J1):@FREE(VARK(J1)));
 @FOR(N(i):@BIN(E(i)));
ENDSUBMODEL
SUBMODEL SVM:
 MIN=ER;
 ER=@SUM(P(J):VARK(J)^2)/2+BIGM*@SUM(N(i):E(i));
 @FOR(N(i):@SUM(P1(J1):IS(i,J1)*VARK(J1)*CHOICE(k,J1)) > 1-E(i));
 @FOR(P1(J1):@FREE(VARK(J1)));
ENDSUBMODEL

SUBMODEL LP: ... ENDSUBMODEL
SUBMODEL IPLP: ...ENDSUBMODEL
SUBMODEL HSVM: ... ENDSUBMODEL
```

The CALC section controls the optimization models by the programming language or makes a report and graph.

## 2.3 Comparison of Seven LDFs

```
CALC:
 @SET('DEFAULT'); @SET('TERSEO',2);

G=1;
K=1; LEND=@SIZE(MS);
@WHILE (K#LE#LEND: @FOR ( P1( J): VARK( J) = 0;
@RELEASE (VARK( J)));
NM=0; NMP=0; Z=0; BIGM=10000;
@SOLVE(RIP);
@FOR(P1(J1): VARK75(@SIZE(MS)*(G-1) + K, J1)=VARK(J1)*CHOICE(k,J1));
VARK75(@SIZE(MS)*(G-1) + K, @SIZE(P2)) = G;
@FOR(n(I): @IFC(SK(I+NN*(G-1), K) # EQ # 0: Z=Z+1));
@FOR(n(I): @IFC(SK(I+NN*(G-1), K) # LT # 0: NM = NM+1));
@FOR(n(I): @IFC(SK(I+NN*(G-1), K) # GT # 0: NMP = NMP+1));
IC(K,G) = NM; ZERO(K,G)=Z;NP(K, G)=NMP;
K=K+1);

G=2;
K=1;LEND=@SIZE(MS);
@WHILE(K#LE#LEND: @FOR( P1( J): VARK( J) = 0;
@RELEASE (VARK( J)));
BIGM=10000; NM=0;NMP=0; Z=0;
@SOLVE(SVM);                  !SVM4;
@FOR(P1(J1):VARK75(@SIZE(MS)*(G-1)+K,J1)=VARK(J1)*CHOICE(k,J1) );
 VARK75(@SIZE(MS)*(G-1)+K,@SIZE(P2))=G ;
@FOR(n(I):SK(I+NN*(G1),K)
   =@SUM(P1(J1):IS(i,J1)*VARK(J1)*CHOICE(k,J1)));
@FOR(n(I):@IFC(SK(I+NN*(G-1),K)#EQ#0: Z=Z+1));
@FOR(n(I):@IFC(SK(I+NN*(G-1),K)#LT#0: NM=NM+1));
@FOR(n(I):@IFC(SK(I+NN*(G-1),K)#GT#0: NMp=NMp+1));
IC(K,G)=NM; ZERO(K,G)=Z;NP(K,G)=NMP;
K=K+1);

!Insert three models from G=3 to G=5;
ENDCALC
```

The DATA section outputs the optimization results to Excel array names, such as "IC, ZERO, NP, and VAR75."

```
DATA:
    @OLE() = IC,ZERO,NP,VARK75;
ENDDATA
END
```

## 2.4 100-Fold Cross-Validation for Small Sample Method (Method 1)

### 2.4.1 Four Trials to Obtain Validation Sample

Until now, we have not been able to discuss the best model and 95 % CI of the error rate and discriminant coefficient for the original data because the discriminant analysis is not the inferential statistical method (Problem 4). We attempt the four approaches described in the following subsections.

#### 2.4.1.1 Generate Training and Validation Samples by Random Number

We create a normal random number data by Speakeasy as follows (Shinmura and Tarumi 2000; Shinmura 2000b):

```
R=ARRAY (400, 2:);      !Array R consists 400 rows by 2 columns;
R=NORMRANDOM (R);       ! We store normal random values with m=0 and SD=1;
R(,1)= R(,1) * 2;       !The SD of first column is two;
```

First, we generate array R with 400 cases (rows) × 2 variables (column) and store the normal random numbers, such as $N(0, 1)$. The first column is variable $x$ and the second column is variable $y$. Both means are zero. The standard deviation (SD) of x is two by the third command and that of y is one. This array is divided into four datasets called "G1, G2, G3, and G4" and consists of 100 cases × two variables. G1 consists of 100 cases from the first to 100th rows. G2 consists of 100 cases from the 101st to 200th rows. G3 consists of 100 cases from the 201st to 300th rows. G4 consists of 100 cases from the 301st to 400th rows. G1 and G2 are used for the training (internal) samples of the discriminant analysis. The remaining G3 and G4 are used as the validation (external) samples. That is, G3 and G4 are the validation sample of G1 and G2, respectively. We create 115 datasets from (G1, G2) and (G3, G4) by the combinations of the rotation for (G1, G2) and the translation for (G3, G4). (G1, G2) are rotated 0° to 30°, 45°, 60°, and 90°, thus setting the center of gravity (means of $x$ and $y$) at the origin. The rotation is achieved by multiplying the rotation matrix by (G1, G2). With regard to (G3, G4), we add integers 0, 1, ..., 8 to $x$ and 0, 2, 4 to $y$. The translation is performed simply as the array operation "G2(, 1) + $i$; G2(, 2) + $j$;" and likewise for G4. Speakeasy is used for this operation. We make 115 datasets by the combination (135 datasets) of five rotations × nine translations of $x$ × three translations of $y$. We omit 20 datasets from the analysis for the following two reasons:

## 2.4 100-Fold Cross-Validation for Small Sample Method (Method 1)

1. If two classes are too separate, the error rates became zero and offer no information with which to evaluate the four discriminant functions, such as IP-OLDF, LP-OLDF, Fisher's LDF, and QDF. We did not understand the importance of LSD in 2000, and thought the discrimination LSD was easy.
2. If two classes are too close, the error rates are close to 0.5 and the discrimination is not meaningful in the actual application. This judgment is correct.

We evaluate four NMs for each discriminant function by the simple regression analysis using MNM as the independent variable in Eq. (2.1). In the training samples, we obtain the following four simple regression lines. If we draw four regression lines, we can determine their ranking by the predicted values as follows: IP-OLDF < QDF < LDF < LP-OLDF. Therefore, QDF is the second best and LP-OLDF is the worst. If we evaluate the four discriminant functions by the random normal data, we forecast that Fisher's LDF and QDF based on the variance–covariance matrices are better than LP-OLDF (Shinmura 1998; 2000a; 2003; 2004; 2005; 2007). However, IP-OLDF is better than Fisher's LDF and QDF.

$$\text{MNM} = 1 + 1 \times \text{MNM} \ (R^2 = 1)$$
$$\text{NM of QDF} = 1.735 + 1.038 \times \text{MNM} \ (R^2 = 0.990)$$
$$\text{NM of LDF} = 2.181 + 1.104 \times \text{MNM} \ (R^2 = 0.992)$$
$$\text{NM of LP-OLDF} = 1.243 + 1.300 \times \text{MNM} \ (R^2 = 0.984)$$
(2.1)

In the validation samples, we obtain the four simple regression lines shown in Eq. (2.2). We can determine the ranking by the predicted values as follows: QDF < IP-OLDF < LDF < LP-OLDF. Therefore, QDF is better than IP-OLDF for the following two reasons:

1. Although QDF has five independent variables, the other three LDFs have two independent variables. In general, the result with more variables is better than the result with fewer variables.
2. Random number data are suitable for QDF and Fisher's LDF based on the variance–covariance matrices. However, we determined that QDF and Fisher's LDF are weak for actual data because most actual data do not satisfy the Fisher's assumption.

$$\text{NM of QDF} = 4.399 + 1.170 \times \text{MNM} \ (R^2 = 0.974)$$
$$\text{NM of IP-OLDF} = 5.430 + 1.229 \times \text{MNM} \ (R^2 = 0.985)$$
$$\text{NM of LDF} = 5.122 + 1.269 \times \text{MNM} \ (R^2 = 0.981)$$
$$\text{NM of LP-OLDF} = 4.594 + 1.510 \times \text{MNM} \ (R^2 = 0.984)$$
(2.2)

### 2.4.1.2 20,000 Normal Random Sampling

We compute the averages and variance–covariance matrices of two classes and generate 20,000 normal random samples with "Monako[3]" in order to create the random sample only once. We use this random sample as a validation sample and compare the results of the original data (training sample). In the validation of the original Iris data and random sample, we accept almost the same results as Eq. (2.2). In the validation of CPD data, we obtain an error rate for the validation sample that is often less than that of the original sample because the validation sample fits better the variance–covariance matrices computed from the original data compared with the original data.

### 2.4.1.3 20,000 Resampling Samples

We generate 20,000 resampling samples from the original data. We are successful in the validation of the result from the original data. However, we cannot obtain useful results, such as the best model and 95 % CI of the error rate and discriminant coefficient.

### 2.4.1.4 *K*-Fold Cross-Validation for Small Sample Method

First, we set $k = 10$. However, we realize that this validation sample provides no useful results. Then, we change $k = 100$ and obtain the best model and 95 % CI of the error rate and discriminant coefficient. We could obtain useful results for the best models and could not explain the useful meaning of the 95 % CI of the discriminant coefficient (Shinmura 2010a). Because the distribution of coefficients in the 100 training samples often has a broad range of negative to positive values, we cannot explain the useful meaning of the 100 coefficients. In 2015, we normalized the coefficient by the intercept of six MP-based LDFs, including H-SVM, and compared Fisher's LDF and logistic regression. All LDFs, with the exception of Fisher's LDF, became trivial LDFs (Shinmura 2015b). We developed IP-OLDF first in our research and set the intercept of IP-OLDF to one. We regret not adopting this idea for the coefficient.

## 2.4.2 Best Model Comparison

Table 2.7 lists the results of 15 models by Method 1. The first column, "RIP," shows all possible combination models from a four to one-variable model, as shown in column "Model." "M1 and M2" columns are the error rate means from the

---

[3]Japanese software.

## 2.4 100-Fold Cross-Validation for Small Sample Method (Method 1)

**Table 2.7** Best Models of seven LDFs (Bold figures are seven best models and minimum Diffs)

| RIP 12m11s | M1 | M2 | Diff. | Model |
|---|---|---|---|---|
| 1 | 0.56 | **2.72** | 2.16 | X1, X2, X3, X4 |
| 2 | 0.96 | 3.03 | 2.07 | X2, X3, X4 |
| 3 | 1.37 | 3.42 | 2.05 | X1, X3, X4 |
| 4 | 2.68 | 5.07 | 2.39 | X1, X2, X4 |
| 5 | 1.55 | 3.70 | 2.15 | X1, X2, X3 |
| 6 | 3.61 | 5.79 | 2.18 | X2, X4 |
| 7 | 2.44 | 4.39 | 1.95 | X3, X4 |
| 8 | 2.91 | 4.82 | 1.91 | X1, X3 |
| 9 | 4.23 | 5.69 | 1.46 | X1, X4 |
| 10 | 4.29 | 7.03 | 2.74 | X2, X3 |
| 11 | 22.74 | 27.27 | 4.53 | X1, X2 |
| 12 | 5.40 | 6.08 | **0.68** | X4 |
| 13 | 5.88 | 7.25 | 1.37 | X3 |
| 14 | 25.75 | 28.24 | 2.49 | X1 |
| 15 | 35.67 | 38.93 | 3.26 | X2 |

| SVM4 8m43s | M1 | M2 | Diff. | M1Diff. | M2Diff. |
|---|---|---|---|---|---|
| 1 | 1.21 | **3.03** | 1.82 | 0.65 | 0.31 |
| 12 | 6.00 | 6.06 | **0.06** | 0.60 | −0.02 |

| SVM1 8m42s | M1 | M2 | Diff. | M1Diff. | M2Diff. |
|---|---|---|---|---|---|
| 1 | 2.23 | **3.00** | 0.77 | 1.67 | 0.28 |
| 12 | 6.16 | 6.28 | **0.12** | 0.76 | 0.20 |

| LP 4m20s | M1 | M2 | Diff. | M1Diff. | M2Diff. |
|---|---|---|---|---|---|
| 1 | 1.15 | **2.98** | 1.83 | 0.59 | 0.26 |
| 12 | 5.74 | 5.83 | **0.09** | 0.34 | −0.25 |

| IPLP 16m39s | M1 | M2 | Diff. | M1Diff. | M2Diff. |
|---|---|---|---|---|---|
| 1 | 0.56 | **2.70** | 2.14 | 0.00 | −0.02 |
| 12 | 5.44 | 6.08 | **0.64** | 0.04 | 0.00 |

| Logistic 18m | M1 | M2 | Diff. | M1Diff. | M2Diff. |
|---|---|---|---|---|---|
| 1 | 1.36 | **3.07** | 1.71 | 1.50 | 0.35 |
| 15 | 40.68 | 40.30 | **−0.38** | 5.01 | 1.37 |

| LDF 16m | M1 | M2 | Diff. | M1Diff. | M2Diff. |
|---|---|---|---|---|---|
| 1 | 2.76 | **3.18** | 0.42 | 2.20 | 0.46 |
| 15 | 40.72 | 40.30 | **−0.42** | 5.05 | 1.37 |

training and validation samples. $M1$ decreases monotonously, similarly to MNM, because $M1$ is the average of 100 MNMs. Therefore, $M1$ of the full model is always the minimum value. We can confirm this fact by the $M1$ values in the table. Although $M2$ of the full model is the minimum value, and it is 2.72, this might be because the data have only four variables. We consider the model with minimum $M2$ to be the best model. We claim that the best model has good generalization ability. Column "Diff" is the difference defined as $(M2 - M1)$. Some statisticians erroneously believe that the model with a minimum value of "Diff" has good generalization ability. Because a one-variable model (X4) has a minimum "Diff" value, these statistics are not useful for selecting the best model. We confirmed this fact through many datasets. We summarize 15 models of other LDFs in two rows. The first row corresponds to the full model. All LDFs select the full model as their best models. Those M2s are 3.03, 3.00, 2.98, 2.70, 3.07, and 3.18 %. The second row corresponds to the model with minimum "Diff." The last two columns, "M1Diff & M2Diff," are the difference defined as (M1/M2 of other LDFs − M1/M2 of RIP). If we focus on "M2Diff" of the full model, those are 0.31, 0.28, 0.26, −0.02, 0.35, and 0.46 % higher than Revised IP-OLDF. Therefore, six LDFs are more acceptable than Revised IP-OLDF. The values of "M2Diff" are almost less than those of "M1Diff." This fact could imply that Revised IP-OLDF over fits the training sample. We observed this defect only with Iris data. If we check "Diff," we can understand that this claim is not correct. In particular, although "Diff" of Fisher's LDF is −0.42 %, this result is caused by a high value of M1, such as 40.72 %. We claim that the full model of Revised IPLP-OLDF has good generalization ability among eight LDFs. The CPU times listed in the first row indicate that Fisher's LDF and logistic regression are slower than MP-based LDFs.

### 2.4.3 Comparison of Discriminant Coefficient

Table 2.8 lists the three percentiles of the discriminant coefficients and intercept. To set the intercept to one, we divide the original five coefficients by the value of (original intercept + 0.00001) in order to avoid division by zero if the original intercept is equal to zero. By setting the intercept to one, we can understand the meaning of the 95 % CI of coefficients clearly. Before normalizing the intercept, we struggle many 95 % CI of coefficients ranging from negative to positive values and include 0 because the intercept almost has both positive and negative values (Shinmura 2015a). Although we proposed this idea, we could not obtain good results because we did not set the intercept to one. Four 95 % CIs of the full model of Revised IP-OLDF include zero, and we cannot reject the null hypothesis at the 5 % level. On the other hand, we can reject three coefficients from a three-variable model (X2, X3, X4) at the 5 % level. Although the 95 % CI recommends the

Table 2.8 95% CI of LDFs (Bold figures show three coefficients are rejected at 5 % level)

| | % | X1 | X2 | X3 | X4 | C |
|---|---|---|---|---|---|---|
| RIP | 97.5 | 4.55 | 5.35 | 9.94 | 12.31 | 1 |
| | 50 | 0.06 | 0.11 | −0.23 | −0.41 | 1 |
| | 2.5 | −5.59 | −11.94 | −6.93 | −6.34 | 1 |
| | 97.5 | | **1.25** | **−0.06** | **−0.14** | 1 |
| | 50 | | **0.18** | **−0.15** | **−0.54** | 1 |
| | 2.5 | | **0** | **−0.53** | **−1.36** | 1 |

## 2.4 100-Fold Cross-Validation for Small Sample Method (Method 1)

three-variable model, we judge the full model selected by the best model is better than three-variable model without theoretical explanation in this book.

If we select the medians as coefficients, we obtain the Revised IP-OLDF in Eq. (2.3). Although we determine that the full model of Revised IP-OLDF is the best model, the 95 % CI of Revised IP-OLDF indicates that this model might be redundant and suggest a three-variable model as a useful model. There is a mismatch between our judgment of the model selection using $M2$ and the 95 % CI of discriminant coefficients in the best model. We usually experience this uncertainty in inferential statistics.

$$\text{RIP} = 0.18 \times X2 - 0.15 \times X3 - 0.54 \times X4 + 1 \tag{2.3}$$

We cannot reject four coefficients of Revised LP-OLDF in Eq. (2.4), three coefficients of Revised IPLP-OLDF in Eq. (2.5), and two coefficients of SVM4 in Eq. (2.6). We can only reject four coefficients of SVM1 in Eq. (2.7). If we check a three-variable model, we can reject three coefficients of four LDFs, similar to Revised IP-OLDF. Before we set the intercept to one, we lost much research time and had no knowledge about discriminant coefficients. However, we cannot obtain clear results from the 95 % CI of the coefficient. On the other hand, we obtain clear results from the exam scores in Chap. 5, the Swiss banknote data (Flury and Rieduyl 1988) in Chap. 6, and Japanese-automobile data in Chap. 7 because these data are LSD.

$$\begin{array}{c} \text{LP} = 0.06 \times X1 + 0.13 \times X2 - 0.21 \times X3 - 0.46 \times X4 + 1 \\ [-0.3, 1.8], \ [-0.7, 2], \ [-2.6, 0.5], \ [-2.8, 0.4] \end{array} \tag{2.4}$$

$$\begin{array}{c} \text{IPLP} = 0.52 \times X1 + 0.11 \times X2 - 0.21 \times X3 - 0.39 \times X4 + 1 \\ [-0.1, 1.6], \ [-0.1, 2.3], \ [-2.6, 0.1], \ [-2.7, -0.1] \end{array} \tag{2.5}$$

$$\begin{array}{c} \text{SVM4} = 0.06 \times X1 + 0.13 \times X2 - 0.22 \times X3 - 0.43 \times X4 + 1 \\ [-0.1, 0.7], \ [-0.1, 0.8], \ [-1.1, -0.03], \ [-1.2, -0.1] \end{array} \tag{2.6}$$

$$\begin{array}{c} \text{SVM1} = 0.08 \times X1 + 0.11 \times X2 - 0.28 \times X3 - 0.28 \times X4 + 1 \\ [0.02, 0.2], \ [0.01, 0.3], \ [-0.5, -0.2], \ [-0.6, -0.2] \end{array} \tag{2.7}$$

## 2.5 Summary

In this chapter, we discussed the Method 1 and the model selection procedure of discriminant analysis using Iris data. Iris data are critical evaluation data used by Fisher until now. Because there are small differences between Fisher's LDF and other LDFs, we should no longer use Iris data as the evaluation data.

# References

Anderson E (1945) The irises of the Gaspe Peninsula. Bull Am Iris Soc 59:2–5
Cox DR (1958) The regression analysis of binary sequences (with discussion). J Roy Stat Soc B 20:215–242
Fisher RA (1936) The use of multiple measurements in taxonomic problems. Annal Eugenics 7:179–188
Fisher RA (1956) Statistical methods and statistical inference. Hafner Publishing Co, New Zealand
Flury B, Rieduyl H (1988) Multivariate Statistics: A Practical Approach. Cambridge University Press
Friedman JH (1989) Regularized discriminant analysis. J Am Stat Assoc 84(405):165–175
Goodnight JH (1978) SAS technical report—the sweep operator: its importance in statistical computing—(R100). SAS Institute Inc, USA
Lachenbruch PA, Mickey MR (1968) Estimation of error rates in discriminant analysis. Technometrics 10:1–11
Miyake A, Shinmura S (1979) An algorithm for the optimal linear discriminant functions. Proceedings of the international conference on cybernetics and society, pp 1447–1450
Miyake A, Shinmura S (1980) An algorithm for the optimal linear discriminant function and its application. Jpn Soc Med Electron Biol Eng 18(6):452–454
Sall JP, Creighton L, Lehman A (2004) JMP start statistics, third edition. SAS Institute Inc., USA. (Shinmura S. edits Japanese version)
Schrage L (2006) Optimization modeling with LINGO. LINDO Systems Inc., USA. (Shinmura S. translates Japanese version)
Shinmura S, Iida K, Maruyama C (1987) Estimation of the effectiveness of cancer treatment by SSM using a null hypothesis model. Inf Health Social Care 7(3):263–275. doi:10.3109/14639238709010089
Shinmura S (1998) Optimal linear discriminant functions using mathematical programming. J Jpn Soc Comput Stat 11(2):89–101
Shinmura S, Tarumi T (2000) Evaluation of the optimal linear discriminant functions using integer programming (IP-OLDF) for the normal random data. J Jpn Soc Comput Stat 12(2):107–123
Shinmura S (2000a) A new algorithm of the linear discriminant function using integer programming. New Trends Prob Stat 5:133–142
Shinmura S (2000b) Optimal linear discriminant function using mathematical programming. Dissertation, 200:1–101. Okayama University, Japan
Shinmura S (2003) Enhanced algorithm of IP-OLDF. ISI2003 CD-ROM, pp 428–429
Shinmura S (2004) New algorithm of discriminant analysis using integer programming. IPSI 2004 Pescara VIP Conference CD-ROM, pp 1–18
Shinmura S (2005) New age of discriminant analysis by IP-OLDF—beyond Fisher's linear discriminant function. ISI2005, pp 1–2
Shinmura S (2007) Overviews of discriminant function by mathematical programming. J Jpn Soc Comput Stat 20(1–2):59–94
Shinmura S (2010a) The optimal linearly discriminant function (Saiteki Senkei Hanbetu Kansuu). Union of Japanese Scientist and Engineer Publishing, Japan
Shinmura S (2010b) Improvement of CPU time of revised IP-OLDF using linear programming. J Jpn Soc Comput Stat 22(1):39–57
Shinmura S (2011a) Beyond Fisher's linear discriminant analysis—new world of the discriminant analysis. ISI CD-ROM, pp 1–6
Shinmura S (2011b) Problems of discriminant analysis by mark sense test data. Jpn Soc Appl Stat 40(3):157–172
Shinmura S (2013) Evaluation of optimal linear discriminant function by 100-fold cross-validation. ISI2013 CD-ROM, pp 1–6
Shinmura S (2014a) End of discriminant functions based on variance-covariance matrices. ICORE2014, pp 5–16

Shinmura S (2014b) Improvement of CPU time of linear discriminant functions based on MNM criterion by IP. Stat Optim Inf Comput 2:114–129

Shinmura S (2014c) Comparison of linear discriminant functions by $K$-fold cross-validation. Data Anal 2014:1–6

Shinmura S (2015a) The 95 % confidence intervals of error rates and discriminant coefficients. Stat Optim Inf Comput 2:66–78

Shinmura S (2015b) A trivial linear discriminant function. Stat Optim Inf Comput 3:322–335. doi:10.19139/soic.20151202

Shinmura S (2015c) Four serious problems and new facts of the discriminant analysis. In: Pinson E, Valente F, Vitoriano B (ed) Operations research and enterprise systems, pp 15–30. Springer, Berlin (ISSN: 1865-0929, ISBN: 978-3-319-17508-9, doi:10.1007/978-3-319-17509-6)

Shinmura S (2015d) Four problems of the discriminant analysis. ISI 2015:1–6

Shinmura S (2016a) The best model of Swiss banknote data. Stat Optim Inf Comput, 4:118–131. International Academic Press (ISSN: 2310-5070 (online) ISSN: 2311-004X (print), doi:10.19139/soic.v4i2.178)

Shinmura S (2016b) Matroska feature selection method for microarray data. Biotechnology 2016:1–8

Shinmura S (2016c) Discriminant analysis of the linear separable data—Japanese automobiles. J Stat Sci Appl X, X:0–14

Vapnik V (1995) The nature of statistical learning theory. Springer, Berlin

# Chapter 3
# Cephalo-Pelvic Disproportion Data with Collinearities

## 3.1 Introduction

In this chapter, we discriminate cephalo-pelvic disproportion (CPD) data (Miyake and Shinmura 1980). These data have significant relationships with the Theory.

1. We evaluated a heuristic optimal linear discriminant function (OLDF) by these data (Miyake and Shinmura 1979; Shinmura and Miyake 1979). However, we could only evaluate a six-variable model because our CPU power was poor on an IBM 360 and because of the limitations of a heuristic OLDF. Therefore, we could not extend our research.
2. These data consist of 240 patients with 19 independent variables. We specified three collinearities in these data and established how to remove such collinearities (Shinmura 1998).
3. We found a strange trend of NMs by quadratic discriminant function (QDF) and found that QDF is fragile for collinearities. Moreover, NM of Fisher's LDF did not decrease in the 19 models from the one-variable model to the 19-variable model selected by the forward and backward stepwise procedure. On the other hand, NMs of our three MP-based optimal LDFs (OLDFs) almost decreased (Shinmura and Tarumi 2000; Shinmura 2000a, b, 2003, 2004, 2011a).
4. In the CPD data, we selected a four-variable model as useful for the regression model selection procedure (plug-in rule1). However, the new model selection procedure that uses the 100-fold cross-validation for small sample method (Method 1) recommends a nine-variable model as the best model. This procedure is simple and powerful compared with leave-one-out (LOO) procedure (Lachenbruch and Mickey 1968). We believe that many variables and/or collinearities cause this difference. Because the Iris data (Anderson 1945) have four variables and might satisfy Fisher's assumption, the model selection procedure by regression analysis and the best model select the full model of seven LDFs, which are Revised IP-OLDF (Shinmura 2005, 2007, 2010a), Revised LP-OLDF, Revised IPLP-OLDF (Shinmura 2010b, 2014b), Support Vector Machine (SVM4

and SVM1) (Vapniik 1995), Fisher's LDF (Fisher 1936, 1956), and logistic regression (Cox 1958; Firth 1993). Because there are slight differences between the seven error rate (Miyake and Shinmura 1976) means from the validation samples (M2), we should no longer use the Iris data as the evaluation data.
5. CPD data has many OCPs. This fact implies that Revised IP-OLDF might search for several OCPs with the same minimum number of misclassifications (MNMs), and different coefficients groups that belong to different OCPs. This result means that it is difficult for us to evaluate the 95 % CI of discriminant coefficients. This is the new Problem 6.

In this chapter, we solve the aforementioned problems.

## 3.2 CPD Data

### 3.2.1 Collinearities

Prof. Suzumura from Nihon Medical School developed Suzumura's method in order to determine which treatments a given surgeon should select for pregnant women with cephalo-pelvic disproportion symptom: Cesarean section or natural delivery. His method is as follows: First, a copy of the fetus' head is made on paper from the X-ray image of the pelvis and cut off. Next, the paper is moved back to the X-ray in order to determine the correct treatment. Prof. Suzumura's staff collected CPD data, and we analyzed these data in order to prove the validity of his method (Miyake and Shinmura 1980). The data consisted of two classes: 180 women who delivered naturally and 60 who delivered by Cesarean section. The staff collected the 19 variables listed in Table 3.1. Three variables ($X1$–$X3$) are standard items. Six variables ($X4$–$X9$) are measurements of the side of the uterus. Five variables ($X10$–$X14$) are measurements of the pelvic inlet image. Five variables ($X15$–$X19$) are measurements of the outside of the uterus.

When we discriminate these data by Revised IP-OLDF, LINGO (Schrage 1991, 2006) indicates that there are two linear relationships among six variables in Eq. (3.1) as a warning error. Therefore, we add slight random noise to $X9$ and $X12$. If we could prove the importance of $X9$ and $X12$, we can consider that Suzumura's method is meaningful for medical diagnosis. However, these variables caused collinearity because the two variables are the subtraction of another variable. Collinearities mean that there are linear relationships of independent variables.

$$\begin{aligned} X9 &= X7 - X8 \\ X12 &= X13 - X14 \end{aligned} \quad (3.1)$$

Some researchers erroneously believe that high correlation among independent variables causes collinearities. Table 3.2 lists the top ten high correlations. The two correlations of ($X7$, $X9$) and ($X12$, $X13$) including Eq. (3.1) are 0.89 and 0.78,

## 3.2 CPD Data

**Table 3.1** Nineteen variables and VIFs

|  | Var. | Description | VIF |
|---|---|---|---|
| Standard items | $X1$ | Age of the pregnant woman | 1.2 |
|  | $X2$ | Number of times of delivery | 1.3 |
|  | $X3$ | Number of sacrum | 1.1 |
| Measurements on uterus side | $X4$ | Utero-posterior distance at pelvic inlet | **24.6** |
|  | $X5$ | Utero-posterior distance at wide pelvis | 8.7 |
|  | $X6$ | Utero-posterior distance at narrow pelvis | 3.1 |
|  | $X7$ | Shortest anteroposterior distance | **57.0** |
|  | $X8$ | Biparietal fetal diameter | 5.3 |
|  | $X9$ | $X9 = X7 - X8$ | **21.0** |
| Measurements of pelvic inlet image | $X10$ | Utero-posterior distance at pelvic inlet | 3.7 |
|  | $X11$ | Biparietal diameter at pelvic inlet | 1.7 |
|  | $X12$ | $X12 = X13 - X14$ | **1484** |
|  | $X13$ | Area at pelvic inlet | **1466** |
|  | $X14$ | Area of fetal head | **638** |
| Measurements of outside of uterus | $X15$ | Area at bottom length of uterus | 1.4 |
|  | $X16$ | Abdominal circumference | 1.7 |
|  | $X17$ | External conjugate | 1.6 |
|  | $X18$ | Intertrochanteric diameter | 1.6 |
|  | $X19$ | Lateral conjugate | 1.4 |

**Table 3.2** Top ten correlations

| Var1. | Var2. | Correlation | Lower 95 % | Upper 95 % | $p$ value |
|---|---|---|---|---|---|
| $X4$ | $X7$ | 0.97 | 0.96 | 0.98 | 3.4E−147 |
| $X7$ | $X9$ | **0.89** | 0.86 | 0.92 | 2.13E−84 |
| $X4$ | $X9$ | 0.87 | 0.83 | 0.90 | 3.28E−74 |
| $X5$ | $X7$ | 0.86 | 0.82 | 0.89 | 6.49E−72 |
| $X7$ | $X10$ | 0.81 | 0.76 | 0.85 | 1.15E−56 |
| $X5$ | $X6$ | 0.78 | 0.73 | 0.83 | 3.33E−51 |
| **X12** | **X13** | **0.78** | 0.73 | 0.83 | 3.53E−51 |
| $X5$ | $X10$ | 0.78 | 0.72 | 0.82 | 3.18E−50 |

respectively, and are not a strong correlation. In addition, the previous claim does not have a correlation threshold for collinearity.

Sall, the founder of JMP (Sall et al. 2004) and vice president of SAS Institute Inc., explained collinearity clearly by a variance inflation factors (VIF), as indicated in Table 3.1. We translated his technical report (Sall 1981) and published a book with Dr. Goodnight's technical report of the sweep operator about all possible regression models (Goodnight 1978). If we regress one independent variable $x_i$ by other independent variables, we obtain the deterministic coefficient $R_i^2$. VIF is defined in Eq. (3.2). Moreover, tolerance (TOL) is the same statistics in Eq. (3.3). Because the range of TOL is [0, 1] and VIF is greater than one, we should use VIF because the range of it is wider than TOL.

$$\text{VIF}_i = 1/(1 - R_i^2) \tag{3.2}$$

$$\text{TOL}_i = 1 - R_i^2 = 1/\text{VIF} \tag{3.3}$$

Table 3.1 lists VIFs. From our experience of CPD data, we determined collinearity by the condition "VIF ≥ 50;" in other words, $R_i^2 \geq 0.98$ and TOL ≤ 0.02. However, we asked Dr. Sall about the threshold in 2015. He answered that it depends on the data, and it is difficult to set the collinearity threshold uniquely. Therefore, we consider that collinearity might be the condition "VIF ≥ 20" that is equal to $R_i^2 = 0.95$. The six VIFs of X4, X7, X9, X12, X13, and X14 are >20. We think "VIF ≥ 20 or $R_i^2 \geq 0.95$" indicates the collinearity. However, we cannot know the linear relationship among collinearities.

## 3.2.2 How to Find Linear Relationships in Collinearities

Table 3.3 lists the eigenvalues (left) and eigenvectors (right) by PCA. Eigenvalues for the 19th, 18th, 17th, and 16th principal components are 0.00, 0.01, 0.04, and 0.15, respectively, and the contribution rates are 0, 0.07, 0.19, and 0.77 %, respectively. These four principal components explain the 0.25 % of total variance of data. If we focus on the large values of eigenvectors that are greater than 0.1, we can drive four linear relationships of collinearities in Eq. (3.4). For example, the 19th principal component is expressed by three variables, such as (X12, X13, X14). If we assign the value of some cases to this formula, the value of the 19th principal component can be calculated. However, its variance is almost zero. Although the variances of the 18th, 17th, and 16th principal components are 0.07, 0.19, and 0.77 %, respectively, we can consider that the three variances are almost zero. Therefore, we find four linear relationships that present collinearities. Next, we can determine the variables that should be removed from the 19 variables in order to eliminate collinearity by the stepwise method.

$$\begin{aligned}
\text{PRIN19:}\ & 0.001 \times X1 - 0.003 \times X2 + 0.001 \times X3 + 0.000 \times X4 - 0.004 \times X5 + 0.000 \times X6 \\
& + 0.002 \times X7 + 0.000 \times X8 + 0.001 \times X9 + 0.002 \times X10 - 0.001 \times X11 \\
& + 0.643 \times X12 - 0.639 \times X13 + 0.422 \times X14 - 0.001 \times X15 + 0.001 \times X16 \quad (3.4) \\
& + 0.001 \times X17 + 0.000 \times X18 + 0.000 \times X19 \\
& = 0.643 \times X12 - 0.639 \times X13 + 0.422 \times X14 = 0 \quad (0\%)
\end{aligned}$$

## 3.2 CPD Data

**Table 3.3** Eigenvalues (*left*) and eigenvectors (*right*)

| Eigenvalues ||||| Eigenvectors |||||
|---|---|---|---|---|---|---|---|---|
| Prin. | Eigen | Contribution | Cum. Cont. | Var. | Prin16 | Prin17 | Prin18 | Prin19 |
| 1 | 6.90 | 36.29 | 36.29 | X1 | 0.015 | 0.004 | 0.001 | 0.001 |
| 2 | 2.14 | 11.26 | 47.55 | X2 | 0.011 | 0.006 | −0.007 | −0.003 |
| 3 | 1.54 | 8.13 | 55.68 | X3 | **0.106** | 0.005 | −0.002 | 0.001 |
| 4 | 1.24 | 6.54 | 62.22 | X4 | *−0.295* | *−0.583* | *0.413* | 0.000 |
| 5 | 1.12 | 5.88 | 68.10 | X5 | *0.795* | *−0.262* | *0.156* | −0.004 |
| 6 | 1.01 | 5.32 | 73.42 | X6 | *−0.406* | 0.050 | −0.048 | 0.000 |
| 7 | 0.94 | 4.95 | 78.37 | X7 | 0.058 | −0.049 | *−0.837* | 0.002 |
| 8 | 0.73 | 3.84 | 82.21 | X8 | 0.045 | *0.303* | *0.134* | 0.000 |
| 9 | 0.62 | 3.28 | 85.50 | X9 | 0.092 | *0.698* | *0.289* | 0.001 |
| 10 | 0.56 | 2.94 | 88.44 | X10 | *−0.277* | 0.062 | 0.020 | 0.002 |
| 11 | 0.52 | 2.76 | 91.20 | X11 | 0.016 | 0.039 | −0.012 | −0.001 |
| 12 | 0.44 | 2.32 | 93.52 | X12 | −0.049 | 0.011 | 0.006 | *0.643* |
| 13 | 0.41 | 2.17 | 95.69 | X13 | −0.026 | −0.002 | 0.002 | *−0.639* |
| 14 | 0.37 | 1.96 | 97.65 | X14 | 0.044 | −0.022 | −0.002 | *0.422* |
| 15 | 0.25 | 1.33 | 98.98 | X15 | 0.031 | −0.017 | 0.000 | −0.001 |
| 16 | **0.15** | **0.77** | **99.75** | X16 | 0.043 | 0.019 | 0.012 | 0.001 |
| 17 | **0.04** | **0.19** | 99.93 | X17 | 0.024 | −0.015 | 0.006 | 0.001 |
| 18 | **0.01** | **0.07** | 100.00 | X18 | −0.074 | 0.023 | −0.003 | 0.000 |
| 19 | **0.00** | **0.00** | 100.00 | X19 | 0.017 | −0.010 | −0.005 | 0.000 |

PRIN18: $0.413 \times X4 + 0.156 \times X5 - 0.837 \times X7 + 0.134 \times X8 + 0.289 \times X9 = 0$  (0.07 %)

PRIN17: $-0.583 \times X4 - 0.262 * X5 + 0.303 \times X8 + 0.698 \times X9 = 0$  (0.19 %)

PRIN16: $0.106 \times X3 - 0.295 \times X4 + 0.795 \times X5 - 0.406 \times X6 - 0.277 \times X10 = 0$ (0.77 %)

Table 3.4 lists all possible combinations of regression models. There are ($2^{19} - 1$) models, from one- to 19-variable models. Therefore, we focus only on those models selected by the forward (F) and backward (B) stepwise procedures. In the "F/B" column of this table, the symbol "F" represents the model selected by the forward stepwise procedure and the symbol "B" represents the model selected by the backward stepwise procedure. We sort all models within the same "$p$" (number of independent variables) in descending order by $R$-square ($R^2$). The column "R(rank)" shows the ranking within the same "$p$." AIC selects a seven-variable model as the backward model. BIC and Cp select the two- and three-variable models selected by both stepwise procedures, respectively. The selected models are quite different. Moreover, the three MNMs are 7, 13, and 12. This fact might imply that we cannot trust these statistics for the model selection of the discriminant analysis

**Table 3.4** All possible combinations of CPD data

| P | R | F/B(JMP)[1] | AIC | BIC | $C_p$ | RIP | OCP | JF[2] | JB[3] | F/B(SAS) | SF[4] | SB[4] |
|---|---|---|---|---|---|---|---|---|---|---|---|---|
| 1 | 1 | FB12 | 110 | 120 | 21 | 20 | 1 | **22** | 22 | FB12 | 22 | 22 |
| 2 | 1 | FB9 | 94 | **108** | 5 | **13** | 3 | 20 | **20** | FB9 | 20 | 20 |
| 3 | 1 | FB18 | 93 | 110 | **3** | **12** | 2 | 22 | 22 | FB18 | 22 | 22 |
| 4 | 1 | F15 | 93 | 113 | 3 | 10 | 2 | 18 | | F15 | 18 | |
| | 3 | B2 | 93 | 114 | 3 | 9 | 1 | | 17 | B13 | | 20 |
| 5 | 1 | B1 | 92 | 116 | 3 | 9 | 1 | | 17 | F17 | 18 | |
| | 4 | F2 | 93 | 117 | 3 | 10 | 1 | 16 | | B14 | | 31 |
| 6 | 1 | B7 | 92 | 119 | 2 | 9 | 4 | | 13 | F2 | 16 | |
| | 8 | F1 | 93 | 120 | 3 | 7 | 5 | 15 | | B15 | | 32 |
| 7 | 1 | B5 | **92** | 123 | 2 | **7** | 10 | | 13 | F1 | 15 | |
| | 9 | F7 | 93 | 124 | 4 | 6 | 1 | 9 | | B17 | | 30 |
| 8 | 2 | F5 | 93 | 127 | 3 | 6 | 1 | 9 | | F7 | 9 | |
| | 3 | B13 | 93 | 127 | 3 | 6 | 3 | | 13 | B1 | | 27 |
| 9 | 1 | B14 | 94 | 131 | 4 | 6 | 3 | | 26 | F5 | 9 | |
| | 2 | F17 | 94 | 131 | 4 | 4 | 3 | 9 | | B2 | | 23 |
| 10 | 1 | B17 | 95 | 135 | 5 | 6 | 13 | | 26 | F19 | 8 | |
| | 11 | F13 | 96 | 136 | 6 | 4 | 6 | 12 | | B7 | | 24 |
| 11 | 1 | F14B15 | 96 | 140 | 6 | 4 | 6 | 22 | 22 | F13/B5 | 9 | 22 |
| 12 | 1 | FB10 | 98 | 145 | 8 | 4 | 1 | 21 | 21 | F14/B19 | 21 | 21 |
| 13 | 2 | F16 | 100 | 150 | 9 | 3 | 5 | 20 | | FB4 | 17 | 17 |
| | 4 | B4 | 100 | 150 | 9 | 3 | 2 | | 21 | | | |
| 14 | 1 | B11 | 102 | 155 | 11 | 3 | 10 | | 18 | FB11 | 16 | 16 |
| | 2 | F4 | 102 | 155 | 11 | 3 | 16 | 18 | | | | |
| 15 | 1 | B19 | 104 | 160 | 12 | 2 | 1 | | 17 | FB16 | 17 | 17 |
| | 2 | F11 | 104 | 160 | 12 | 3 | 1 | 17 | | | | |
| 16 | 1 | F19B16 | 106 | 165 | 14 | 2 | 2 | 17 | 17 | FB8 | 21 | 21 |
| 17 | 1 | FB6 | 108 | 171 | 16 | 2 | 8 | 16 | 16 | FB10 | 19 | 19 |
| 18 | 1 | FB3 | 111 | 176 | 18 | 2 | 15 | 15 | 15 | FB6 | 17 | 17 |
| 19 | 1 | FB8 | 113 | 182 | 20 | 2 | 16 | 15 | 15 | FB3 | 16 | 16 |

[1]F: forward; B: backward; Number: Variable; FB12: X12 selected by F & B
[2]JF: NM of QDF by JMP forward; 22: NM of model (X12) by QDF is 22
[3]JB: NM of QDF by JMP backward; 20: NM of model (X9, X12) by QDF is 20

(Nomura and Shinmura 1978; Shimizu et al. 1975; Shinmura et al. 1983, 1987; Shinmura 2001). The column "OCP" is the number of OCPs. LINGO *K*-best option can find this number, as shown in Fig. 4.3. All models of the Iris, Swiss banknote (Flury and Rieduyl 1988), pass/fail determination (Shinmura 2011b), and Japanese-automobile data (Shinmura 2016) have one "OCP." Although there is no collinearity in Student data (Shinmura 2010a), two models of Student data have many OCPs. Until now, we have used the model sequences selected by the stepwise procedures of SAS calculated before 1981 (Shinmura 1998) in column SF[4] and SB[4].

## 3.2 CPD Data

In 2015, we discriminated CPD data by JMP and obtained several different results in column JF[3] and JB[3].

We compare NMs of QDF by JMP and SAS. From 35 years ago, we used the output by SAS. Forward stepwise of JMP selects a variable from the one- to 19-variable models in column "F/B(JMP) and JF" as follows:

$\underline{X12(22)}$ -> $X9(20)$ -> $X18(22)$ -> $X15(18)$ -> $X2(16)$ -> $X1(15)$ -> $X7(9)$ -> $X5(9)$ -> $X17(9)$ -> $\underline{X13(12)}$ -> $\underline{X14(22)}$ -> $X10(21)$ -> $X16(20)$ -> $X4(18)$ -> $X11(17)$ -> $X19(17)$ -> $X6(16)$ -> $X3(15)$ -> $X8(15)$.

The blanket numbers are NMs of QDF. The underlined expressions indicate three variables that are strong collinearities. After $X14$ is entered into the 11-variable model, NM increases to 22 from 12 because this model includes ($X12$, $X13$, $X14$) and has collinearity. Backward stepwise deletes variables from 19- to one-variable models in column "F/B(JMP) and JB" as follows:

$X8(15)$ -> $X3(15)$ -> $X6(16)$ -> $X16(17)$ -> $X19(17)$ -> $X11(18)$ -> $X4(21)$ -> $X10(21)$ -> $X15(22)$ -> $X17(26)$ -> $\underline{X14(26)}$ -> $\underline{X13(13)}$ -> $X5(13)$ -> $X7(13)$ -> $X1(17)$ -> $X2(17)$ -> $X18(22)$ -> $X9(20)$ -> $\underline{X12(22)}$.

After $X14$ is excluded from the nine-variable model, NM decreases from 26 to 13 in the eight-variable model. Figure 3.1 shows NMs of JMP. The symbol "O" is NM for forward stepwise. NM decreases from the one- to six-variable models. NMs for three models (seven, eight, and nine variables) are nine. After the ten-variable model, NM increases to the 13-variable model. In particular, when $X14$ is entered into the 11-variable model, NM "jumps" to 22 from 12. The symbol "+" is NM for backward stepwise. NM increases from the 19-variable model to the nine-variable model. After $X14$ is excluded from the nine-variable model, NM in the eight-variable model decreases to 13 from 26. The strong relationship of collinearity causes this result mainly.

Figure 3.2 shows NMs of SAS. The forward stepwise procedure selects from one- to 19- variables models in column "F/B(SAS) and SF" as follows:

$\underline{X12(22)}$ -> $X9(20)$ -> $X18(22)$ -> $X15(18)$ -> $X17(18)$ -> $X2(16)$ -> $X1(15)$ -> $X7(9)$ -> $X5(9)$ -> $X19(8)$ -> $\underline{X13(9)}$ -> $\underline{X14(21)}$ -> $X4(17)$ -> $X11(16)$ -> $X16(17)$ -> $X8(21)$ -> $X10(19)$ -> $X6(17)$ -> $X3(16)$.

**Fig. 3.1** NMs of JMP

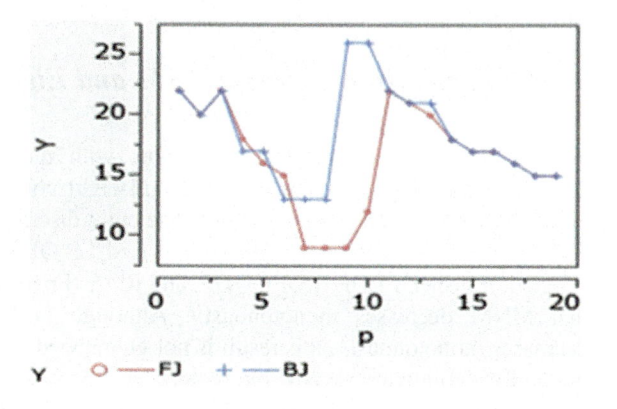

**Fig. 3.2** NMs of SAS

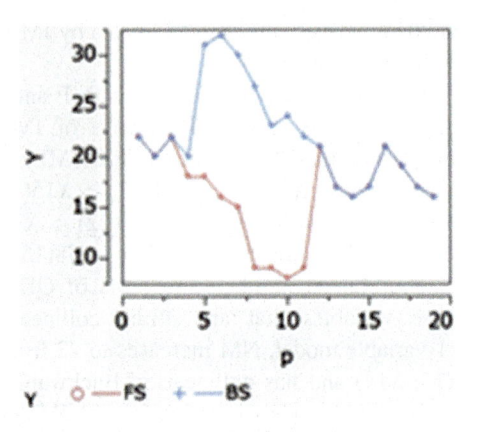

Backward stepwise deletes variables from 19- to one-variable models in column "F/B(SAS) and SB" as follows:

X3(16) -> X6(17) -> X10(19) -> X8(21) -> X16(17) ->-> X11(16) -> X4(17) -> X19(21) -> X5(22) -> X7(24) -> X2(23) -> X1(27) ->-> X17(30) -> X15(32) -> X14(31) -> X13(20) -> X18(22) -> X9(20) -> X12(22).

The models selected by JMP and SAS are different. This might be the result of the computational difficulty of collinearity and/or algorithm change. The symbol "O" is NM of SAS forward stepwise. NM decreases from one to seven variables. NMs of three models (eight, nine, and 11 variables) are nine. NM of ten variables is eight. NM of 12 variables is 21. The symbol "+" is NM of SAS backward stepwise. NM increases from 19 to five variables. After X14 is excluded from the five-variable model, NM decreases. For long-term research, we should register the analysis date and software version because we have had many results that have been different because of software updates. We are concerned that those researchers who follow our papers might become confused. JMP is version 10 and LINGO was version 15 in 2014. However, we have no record on SAS. We analyzed the CPD data on an IBM 360 Model 135 before 1981. We regret having to introduce this old output.

## 3.2.3 Comparison Between MNM and Eight NMs

We compare nine discriminant functions, with the exception of H-SVM. In Table 3.5, the "RIP" column is MNM and eight columns are "Diff1s" (NMs of eight discriminant functions – MNM). The eight discriminant functions are SVM4, SVM1, Revised IPLP-OLDF (IPLP), Revised LP-OLDF (LP), logistic regression (Logistic), Fisher's LDF (LDF), QDF, and RDA (Friedman 1989). We can confirm that MNM decreases monotonously. Although NM of Revised IPLP-OLDF decreases monotonously, this result is not guaranteed theoretically. From the table, we find the following remarkable facts:

## 3.2 CPD Data

**Table 3.5** MNM and eight Diff1s

| p | RIP | SVM4 | SVM1 | IPLP | LP | Logistic | LDF | QDF | RDA |
|---|-----|------|------|------|----|----|-----|-----|-----|
| 1 | 20 | *1* | *1* | 0 | **0** | 6 | 3 | 2 | 2 |
| 2 | 13 | **4** | **4** | 0 | 4 | 16 | **4** | 7 | 7 |
| 3 | 12 | **6** | **6** | 0 | **6** | 12 | 7 | 10 | 9 |
| 4 | 10 | 3 | 3 | 0 | 3 | 8 | 7 | 8 | 8 |
|   | 9 | 6 | 7 | 0 | 6 | 7 | **6** | 8 | 8 |
| 5 | 9 | 6 | 7 | 0 | 6 | 14 | 8 | 8 | 7 |
|   | 7 | 6 | **5** | 0 | 6 | 9 | 10 | 9 | 8 |
| 6 | 9 | 3 | 5 | 0 | 3 | 7 | 6 | 4 | 6 |
|   | 7 | 7 | **5** | 0 | 7 | 7 | 10 | 8 | **5** |
| 7 | 7 | 9 | 10 | 0 | 9 | 8 | 8 | **6** | 8 |
|   | 6 | 6 | 6 | 0 | 6 | 9 | 10 | **3** | 7 |
| 8 | 4 | 5 | 7 | 0 | 5 | **3** | 10 | 5 | 8 |
|   | 6 | 9 | 6 | 0 | 9 | 8 | 10 | 7 | 8 |
| 9 | 6 | 5 | 6 | 0 | **5** | 7 | 10 | 20 | 8 |
|   | 4 | **4** | 7 | 0 | **4** | 8 | 11 | 5 | 8 |
| 10 | 6 | 7 | 7 | 0 | 7 | **5** | 9 | 20 | 7 |
|   | 4 | 5 | **4** | 0 | 5 | **4** | 9 | 8 | 9 |
| 11 | 4 | 5 | **4** | 0 | 5 | **4** | 9 | 18 | 11 |
| 12 | 3 | 5 | 5 | 0 | 5 | **5** | 10 | 18 | 11 |
| 13 | 3 | **3** | 6 | 0 | **3** | 6 | 10 | 17 | 9 |
|   | 3 | **3** | 5 | 0 | **3** | 5 | 10 | 18 | 12 |
| 14 | 3 | 4 | 6 | 0 | 4 | 6 | 13 | 15 | 11 |
|   | 3 | 4 | 6 | 0 | 4 | **4** | 9 | 15 | 9 |
| 15 | 2 | 5 | 6 | 0 | 5 | 7 | 13 | 15 | 11 |
|   | 2 | 5 | 6 | 0 | 5 | **5** | 14 | 15 | 10 |
| 16 | 2 | **3** | 5 | 0 | **3** | 4 | 14 | 15 | 11 |
| 17 | 2 | **3** | 6 | 0 | **3** | 7 | 14 | 14 | 10 |
| 18 | 2 | 4 | 5 | 0 | 4 | 7 | 14 | 13 | 10 |
| 19 | 2 | 4 | 5 | 0 | 4 | 7 | 14 | 13 | 9 |

1. Revised IPLP-OLDF is the same as Revised IP-OLDF. Both LDFs are superior to other LDFs.
2. The bold numbers are the minimum values of "Diff1s" of seven LDFs, with the exception of Revised IPLP-OLDF. Three MP-based LDFs have 48 minimum values. Four statistical LDFs have 12 minimum values. In general, we can determine that three MP-based LDFs are superior to four statistical LDFs.

In general, logistic regression is better than other LDFs, with the exception of Revised IP-OLDF and Revised IPLP-OLDF. However, many NMs of logistic regression are not better than other NMs in CPD data. To this point, we have known that collinearity greatly influences QDF. In this research, we are aware that logistic regression might also be weak for collinearity.

## 3.2.4 Comparison of 95 % CI of Discriminant Coefficient

Table 3.6 lists the 95 % CI of 19 and nine-variable coefficients (Shinmura 2013, 2014a, c, 2015a–d, 2016a). Because all coefficients include zero, we determine that all coefficients, with the exception of the intercept, are zero. Although we cannot explain this result positively, we consider that two possibilities might be collinearity and/or many OCPs.

## 3.3 100-Fold Cross-Validation

### 3.3.1 Best Model

Table 3.7 lists the results of 29 models by the Method 1. Revised IP-OLDF and Revised IPLP-OLDF select a forward nine-variable model as the best model. SVM4 and SVM1 select the full models. LP and logistic regression select the 18-variable models. Fisher's LDF selects the backward seven-variable model. To summarize the previous results, we never permit the 19- or 18-variable models selected by four LDFs because the best model of Revised IPLP-OLDF has the minimum $M2$ among seven best models. We compare $M2$s of the forward nine-variable model of seven LDFs,

Table 3.6 95% CI of 19 and nine-variable models

| RIP | 97.5 | 2.5 | 97.5 | 2.5 |
|---|---|---|---|---|
| 1 | 675,569 | −3857 | 0.08 | −0.59 |
| 2 | 3.202 | 0 | 7.63 | −3.46 |
| 3 | 1.577 | −428,472 | | |
| 4 | 16,051 | −62,257 | | |
| 5 | 0.318 | −99,274 | 0.17 | −0.20 |
| 6 | 17,154 | −21,872 | | |
| 7 | 3188 | −31,079 | 0.30 | −0.29 |
| 8 | 451,121 | −0.692 | | |
| 9 | 177,497 | −31,534 | 0.29 | −0.28 |
| 10 | 17,910 | −21,603 | | |
| 11 | 740 | −84,804 | | |
| 12 | 0.1 | −13,002 | 0.05 | −0.02 |
| 13 | 1269 | −1041 | | |
| 14 | 0.095 | −661 | | |
| 15 | 28,580 | −0.035 | 0.04 | −0.08 |
| 16 | 3034 | −2988 | | |
| 17 | 97,137 | −273.07 | 0.52 | −0.11 |
| 18 | 0.125 | −64,072 | 0.16 | −0.09 |
| 19 | 29,394 | −564.17 | | |
| c | 1 | 0 | 1 | 1 |

3.3 100-Fold Cross-Validation

**Table 3.7** Beat models by 100-fold cross-validation

| RIP | M1 | M2 | Diff. | F/B(JMP) | |
|---|---|---|---|---|---|
| 19 | 0.02 | 3.77 | 3.75 | FB8 | |
| 18 | 0.03 | 3.66 | 3.63 | FB3 | |
| 17 | 0.12 | 3.83 | 3.71 | FB6 | |
| 16 | 0.15 | 3.78 | 3.62 | F19B16 | |
| 15 | 0.24 | 3.78 | 3.54 | B19 | |
| 15 | 0.18 | 3.79 | 3.6 | F11 | |
| 14 | 0.47 | 4.01 | 3.54 | B11 | |
| 14 | 0.32 | 3.9 | 3.58 | F4 | |
| 13 | 0.51 | 3.96 | 3.45 | F16 | |
| 13 | 0.53 | 4.08 | 3.55 | B4 | |
| 12 | 0.59 | 3.96 | 3.37 | FB10 | |
| 11 | 0.69 | 3.96 | 3.28 | F14B15 | |
| 10 | 0.7 | 3.77 | 3.07 | F13 | |
| 10 | 1.22 | 4.38 | 3.15 | B17 | |
| 9 | 0.82 | **3.64** | 2.82 | F17 | |
| 9 | 1.43 | 4.43 | 2.99 | B14 | |
| 8 | 1.05 | 3.72 | 2.66 | F5 | |
| 8 | 1.44 | 4.39 | 2.95 | B13 | |
| 7 | 1.94 | 4.77 | 2.83 | F7 | |
| 7 | 1.65 | 4.36 | 2.7 | B5 | |
| 6 | 2.16 | 4.51 | 2.35 | F1 | |
| 6 | 2.63 | 5.23 | 2.6 | B7 | |
| 5 | 2.56 | 4.6 | 2.04 | F2 | |
| 5 | 3.02 | 5.11 | 2.1 | B1 | |
| 4 | 3.58 | 5.68 | 2.1 | F15 | |
| 4 | 3.32 | 5.01 | 1.69 | B2 | |
| 3 | 4.38 | 5.97 | 1.59 | FB18 | |
| 2 | 4.83 | 6.04 | 1.21 | FB9 | |
| 1 | 7.92 | 9.02 | 1.1 | FB12 | |
| SVM4 | M1 | M2 | Diff. | M1Diff. | M2Diff. |
| 19 | 0.06 | 3.85 | 3.79 | 0.042 | 0.08 |
| 9 | 2.25 | **4.48** | 2.23 | 1.438 | **0.85** |
| SVM1 | M1 | M2 | Diff. | M1Diff. | M2Diff. |
| 19 | 1.13 | **4.61** | 3.48 | 1.117 | 0.85 |
| 9 | 2.93 | **4.93** | 2.01 | 2.108 | **1.3** |
| LP | M1 | M2 | Diff. | M1Diff. | M2Diff. |
| 19 | 0.05 | 3.73 | 3.68 | 0.029 | −0.04 |
| 18 | 0.07 | **3.73** | 3.66 | 0.042 | 0.07 |
| 9 | 2.25 | **4.45** | 2.2 | 1.429 | **0.81** |
| IPLP | M1 | M2 | Diff. | M1Diff. | M2Diff. |
| 9 | 0.83 | **3.62** | 2.8 | 0.008 | **−0.02** |
| Logistic | M1 | M2 | Diff. | M1Diff. | M2Diff. |

(continued)

Table 3.7 (continued)

| | | | | | |
|---|---|---|---|---|---|
| 18 | 0.17 | **3.95** | 3.79 | 0.138 | 0.29 |
| 9 | 2.73 | **4.78** | 2.04 | 1.917 | **1.14** |
| LDF | $M1$ | $M2$ | Diff. | $M1$Diff. | $M2$Diff. |
| 9 | 8.91 | **9.88** | 0.97 | 8.096 | **6.24** |
| 7 | 8.73 | **9.38** | **0.66** | 6.783 | 4.62 |

which are 3.64, 4.48, 4.93, 4.45, 3.62, 4.78, and 9.88 %. "$M2$Diff" of six LDFs are 0.85, 1.30, 0.81, −0.02, 1.14, and 6.24 %. Fisher's LDF is poor. The CPU times of logistic regression and Fisher's LDF are 1 and 1 h 10 min, respectively, by our wristwatch. The CPU times of RIP, IPLP, LP, SVM4, and SVM1 are 1 h 57 min, 2 h 17 min, 46 min, 1 h 34 min, and 1 h 39 min, respectively. Only LP is faster than logistic regression and Fisher's LDF. Although Fisher's LDF and logistic regression have always been slower than MP-based LDFs since 2012, we find that this is the reverse in CPD data. Collinearity might be the cause for this reversal. We estimate that four MP-based LDFs, with the exception of Revised LP-OLDF, are slower than Fisher's LDF and logistic regression because of collinearity. We estimate that IP and QP require more time to converge to the optimal solution for data with collinearity. We must examine this prediction in the near future. We cannot explain the mismatch where we selected good models within five variables in the original data and the best model with nine variables by the 100-fold cross-validation method. Some statisticians claim the model with minimum "Diff" such as seven-variable Fisher's LDF has good generalization ability because such model does not overestimate. However, M1 and M2 of this model are very high.

### 3.3.2  95 % CI of Discriminant Coefficient

Table 3.8 lists the percentiles of three models from one to three variables of Revised IP-OLDF. The 95 % CI of the $X12$ coefficient is [−169.4, −0.22]. We can determine that the $X12$ coefficient is negative. If we use the median as the coefficient, we obtain one-variable LDF in Eq. (3.5). The 95 % CIs of $X9$ and $X12$ are [−917.61, 1.47] and [−128.11, −0.05], respectively. Therefore, we can determine

Table 3.8 Percentiles of three models of Revised IP-OLDF

| % | $X9$ | $X12$ | $X18$ |
|---|---|---|---|
| 97.5 | | −0.22 | |
| 50.0 | | −112.35 | |
| 2.5 | | −169.40 | |
| 97.5 | 1.47 | −0.05 | |
| 50.0 | −621.93 | −36.44 | |
| 2.5 | −917.61 | −128.11 | |
| 97.5 | −0.67 | −0.06 | 191.31 |
| 50.0 | −311.77 | −49.32 | 0.52 |
| 2.5 | −987.37 | −104.87 | −256.48 |

## 3.3 100-Fold Cross-Validation

that the coefficient of $X9$ is zero, and that of $X12$ is negative. We obtain a two-variable model in Eq. (3.6). Because there are three OCPs listed in Table 3.4, 100 LDFs from the training samples might correspond to three different OCPs. The 95 % CIs of ($X9, X12, X18$) are [−987.37, −0.67], [−104.87, −0.06], and [−256.48, 191.31], respectively. Therefore, we can determine that the coefficients of $X9$ and $X12$ are negative, and that of $X18$ is zero. We obtain a three-variable model in Eq. (3.7). For four-variable models and above, we can determine that all coefficients are zero. Therefore, we obtain a mismatch in the conclusion between the best model and 95 % CI of coefficients.

$$\text{RIP} = -112.35 \times X12 + 1 \tag{3.5}$$

$$\text{RIP} = -1621.93 \times X9 - 36.44 \times X12 + 1 \tag{3.6}$$

$$\text{RIP} = -311.77 \times X9 - 49.32 \times X12 + 0.52 \times X18 + 1 \tag{3.7}$$

Because many OCPs have the same MNMs, the best models have the same MNM. On the other hand, we have a serious problem with regard to the 95 % CI of discriminant coefficients because many OCPs have different 95 % CI of discriminant coefficients. Moreover, we cannot explain the complex effects of many OCPs and collinearity for the Theory.

## 3.4 Trial to Remove Collinearity

### 3.4.1 Examination by PCA (Alternative 2)

The most popular treatment for collinearities is the use of PCA. We analyze the original CPD data by PCA and obtain 19 principal components as the new variables, as listed in Table 3.9. If we analyze these new variables, we are free from the effect of collinearity. The fifth component is entered into the three-variable model because it is more valuable than the third and fourth components from the perspective of discrimination. In the table, "eigen." indicates eigenvalues; "contribution" is the contribution ratio, and its first component is 36.29 % of the total data variation; "cum." is cumulative contribution, and its first three components explain 55.68 % of the total data variation. Therefore, we can omit the last three components because those explain only 0.25 % of the total data variation. We obtain AIC, BIC, and Cp by regression analysis, and they recommend nine, four, and nine-variable models as useful. On the other hand, Table 3.4 indicates that seven, two, and three-variable models are useful for the original CPD data. If we use 19 components instead of original variables, we observe that the selected model requires more variables than the original data.

Table 3.10 lists MNM ("RIP" column) and eight "Diff1." Although MNM decreases monotonously, MNMs from eight to 14 variables are six. Therefore, we have a good model within eight variables.

**Table 3.9** 19 variables by PCA and VIFs

| p | Prin. | Eigen | Contribution | Cum. | AIC | BIC | Cp |
|---|---|---|---|---|---|---|---|
| 1 | 1 | 6.90 | **36.29** | 36.29 | 177.9 | 188.3 | 105.1 |
| 2 | 2 | 2.14 | 11.26 | 47.55 | 130.1 | 143.8 | 43.0 |
| 3 | 5 | 1.54 | 8.13 | **55.68** | 116.7 | 133.8 | 27.7 |
| 4 | 3 | 1.24 | 6.54 | 62.22 | 104.3 | **124.9** | 14.5 |
| 5 | 13 | 1.12 | 5.88 | 68.10 | 102.2 | 126.0 | 12.2 |
| 6 | 12 | 1.01 | 5.32 | 73.42 | 100.8 | 128.0 | 10.7 |
| 7 | 8 | 0.94 | 4.95 | 78.37 | 99.6 | 130.2 | 9.4 |
| 8 | 7 | 0.73 | 3.84 | 82.21 | 98.6 | 132.4 | 8.3 |
| 9 | 19 | 0.62 | 3.28 | 85.50 | **98.4** | 135.5 | **8.1** |
| 10 | 14 | 0.56 | 2.94 | 88.44 | 98.6 | 139.0 | 8.2 |
| 11 | 10 | 0.52 | 2.76 | 91.20 | 99.1 | 142.7 | 8.5 |
| 12 | 6 | 0.44 | 2.32 | 93.52 | 99.6 | 147.5 | 8.9 |
| 13 | 18 | 0.41 | 2.17 | 95.69 | 100.2 | 150.2 | 9.3 |
| 14 | 9 | 0.37 | 1.96 | 97.65 | 101.7 | 154.9 | 10.6 |
| 15 | 11 | 0.25 | 1.33 | 98.98 | 103.7 | 160.1 | 12.3 |
| 16 | 16 | 0.15 | 0.77 | **99.75** | 105.8 | 165.3 | 14.1 |
| 17 | 17 | 0.04 | 0.18 | 99.93 | 108.1 | 170.7 | 16.0 |
| 18 | 15 | 0.01 | 0.07 | 100.00 | 110.4 | 176.2 | 18.0 |
| 19 | 4 | 0.00 | 0.00 | 100.00 | 112.8 | 181.7 | 20.0 |

**Table 3.10** Comparison of MNM and eight NMs

| p | RIP | SVM4 | SVM1 | LP | IPLP | Logistic | LDF | QDF | RDA |
|---|---|---|---|---|---|---|---|---|---|
| 1 | 30 | 0 | 0 | 0 | 0 | 96 | 2 | 3 | 3 |
| 2 | 14 | 5 | 5 | 5 | 0 | 139 | 4 | 6 | 5 |
| 3 | 12 | 2 | 2 | 2 | 0 | 27 | 3 | 7 | 7 |
| 4 | 10 | 2 | 3 | 2 | 0 | 46 | 6 | 8 | 7 |
| 5 | 10 | 3 | 3 | 3 | 0 | 23 | 5 | 4 | 5 |
| 6 | 9 | 4 | 4 | 4 | 0 | 12 | 6 | 7 | 8 |
| 7 | 8 | 4 | 4 | 4 | 0 | 2 | 6 | 9 | 8 |
| 8 | 6 | 5 | 7 | 5 | 0 | 4 | 11 | 11 | 9 |
| 9 | 6 | 6 | 7 | 6 | 0 | 2 | 11 | 6 | 9 |
| 10 | 6 | 5 | 6 | 5 | 0 | 2 | 12 | 12 | 14 |
| 11 | 6 | 6 | 8 | 6 | 0 | 2 | 10 | 9 | 11 |
| 12 | 6 | 7 | 8 | 7 | 0 | 2 | 10 | 11 | 11 |
| 13 | 6 | 7 | 8 | 7 | 0 | 6 | 7 | 9 | 10 |
| 14 | 6 | 8 | 8 | 8 | 0 | 6 | 12 | 7 | 11 |
| 15 | 5 | 5 | 7 | 5 | 0 | 4 | 12 | 8 | 6 |
| 16 | 3 | 9 | 10 | 9 | 0 | 12 | 14 | 8 | 7 |
| 17 | 3 | 7 | 10 | 7 | 0 | 8 | 14 | 11 | 12 |
| 18 | 3 | 4 | 7 | 4 | 0 | 6 | 13 | 12 | 10 |
| 19 | 2 | 4 | 7 | 4 | 0 | 7 | 14 | 13 | 14 |

## 3.4 Trial to Remove Collinearity

Table 3.11 lists the results of seven LDFs by the Method 1. Six LDFs select the full models as the best models. Only Fisher's LDF select the 15-variable model. Although logistic regression has the minimum M2 among seven best models, it is 0.01 % less than that of Revised IP-OLDF. "*M*1s and *M*2s" of five LDFs are almost the same as those of Revised IP-OLDF. Only *M*2s of Fisher's LDF are over 5.54 % larger than those of Revised IP-OLDF.

**Table 3.11** Results of seven LDFs by Method 1

| RIP | *M*1 | *M*2 | Diff. | Prin. | |
|---|---|---|---|---|---|
| 1 | 12.05 | 13.27 | 1.23 | 1 | |
| 2 | 5.47 | 6.69 | 1.23 | 2 | |
| 3 | 4.37 | 5.81 | 1.44 | 5 | |
| 4 | 3.57 | 5.50 | 1.93 | 3 | |
| 5 | 3.10 | 5.48 | 2.39 | 13 | |
| 6 | 2.75 | 5.21 | 2.46 | 12 | |
| 7 | 1.92 | 4.66 | 2.74 | 8 | |
| 8 | 1.73 | 4.61 | 2.88 | 7 | |
| 9 | 1.54 | 4.68 | 3.14 | 19 | |
| 10 | 1.27 | 4.70 | 3.44 | 14 | |
| 11 | 1.14 | 4.85 | 3.71 | 10 | |
| 12 | 0.98 | 4.89 | 3.92 | 6 | |
| 13 | 0.88 | 4.88 | 4.00 | 18 | |
| 14 | 0.70 | 5.04 | 4.33 | 9 | |
| 15 | 0.45 | 4.58 | 4.13 | 11 | |
| 16 | 0.26 | 4.37 | 4.10 | 16 | |
| 17 | 0.21 | 4.53 | 4.31 | 17 | |
| 18 | 0.13 | 4.63 | 4.50 | 15 | |
| 19 | 0.02 | **4.03** | 4.01 | 4 | |
| SVM4 | *M*1 | *M*2 | Diff. | *M*1Diff. | *M*2Diff. |
| 19 | 0.02 | **4.03** | 5.38 | 0.000 | **0.00** |
| SVM1 | *M*1 | *M*2 | Diff. | *M*1Diff. | *M*2Diff. |
| 19 | 1.92 | **4.59** | 2.67 | 1.904 | **0.56** |
| LP | *M*1 | *M*2 | Diff. | *M*1Diff. | *M*2Diff. |
| 19 | 0.05 | **4.09** | 4.05 | 0.029 | **0.06** |
| IPLP | *M*1 | *M*2 | Diff. | *M*1Diff. | *M*2Diff. |
| 19 | 0.02 | **4.06** | 4.05 | 0.000 | **0.03** |
| Logistic | *M*1 | *M*2 | Diff. | *M*1Diff. | *M*2Diff. |
| 1 | 13.13 | 13.16 | 0.03 | 1.083 | −0.11 |
| 2 | 7.59 | 7.77 | 0.18 | 2.121 | 1.08 |
| 3 | 5.90 | 6.26 | 0.37 | 1.529 | 0.45 |
| 4 | 5.39 | 6.03 | 0.64 | 1.821 | 0.53 |
| 5 | 5.29 | 5.97 | 0.68 | 2.196 | 0.49 |
| 6 | 5.16 | 6.15 | 0.99 | 2.413 | 0.94 |

(continued)

**Table 3.11** (continued)

| | | | | | |
|---|---|---|---|---|---|
| 7 | 4.60 | 5.99 | 1.39 | 2.675 | 1.32 |
| 8 | 4.55 | 6.21 | 1.66 | 2.825 | 1.60 |
| 9 | 4.29 | 6.29 | 1.99 | 2.754 | 1.61 |
| 10 | 4.06 | 6.22 | 2.16 | 2.796 | 1.52 |
| 11 | 3.84 | 6.18 | 2.34 | 2.700 | 1.33 |
| 12 | 3.71 | 6.16 | 2.45 | 2.733 | 1.27 |
| 13 | 3.59 | 6.22 | 2.63 | 2.704 | 1.34 |
| 14 | 3.23 | 6.23 | 3.00 | 2.529 | 1.20 |
| 15 | 2.38 | 5.56 | 3.18 | 1.929 | 0.98 |
| 16 | 1.20 | 4.82 | 3.61 | 0.942 | 0.45 |
| 17 | 1.06 | 4.93 | 3.87 | 0.846 | 0.40 |
| 18 | 0.78 | 4.73 | 3.95 | 0.650 | 0.10 |
| 19 | 0.09 | **4.02** | 3.93 | 0.075 | **−0.01** |
| LDF | M1 | M2 | Diff. | M1Diff. | M2Diff. |
| 1 | 20.29 | 20.19 | −0.10 | 8.246 | 6.92 |
| 2 | 13.28 | 13.40 | 0.13 | 7.808 | 6.71 |
| 3 | 11.91 | 12.15 | 0.24 | 7.542 | 6.34 |
| 4 | 9.10 | 9.59 | 0.49 | 5.525 | 4.08 |
| 5 | 9.70 | 10.44 | 0.74 | 6.604 | 4.96 |
| 6 | 9.79 | 10.67 | 0.88 | 7.046 | 5.46 |
| 7 | 9.33 | 10.32 | 0.99 | 7.404 | 5.66 |
| 8 | 9.46 | 10.55 | 1.09 | 7.733 | 5.94 |
| 9 | 9.12 | 10.31 | 1.19 | 7.583 | 5.63 |
| 10 | 8.98 | 10.29 | 1.31 | 7.713 | 5.58 |
| 11 | 8.27 | 9.76 | 1.49 | 7.125 | 4.90 |
| 12 | 8.02 | 9.56 | 1.54 | 7.042 | 4.67 |
| 13 | 7.75 | 9.31 | 1.56 | 6.867 | 4.43 |
| 14 | 7.59 | 9.34 | 1.75 | 6.883 | 4.30 |
| 15 | 7.37 | **9.19** | 1.83 | 6.917 | **4.62** |
| 16 | 7.44 | 9.36 | 1.92 | 7.179 | 4.99 |
| 17 | 7.47 | 9.44 | 1.97 | 7.254 | 4.91 |
| 18 | 7.47 | 9.61 | 2.14 | 7.338 | 4.98 |
| 19 | 7.35 | **9.57** | 2.22 | 7.333 | **5.54** |

We summarize our examinations as follows:

1. If we analyze the principal components instead of the original variables, we are free from the tedious research task caused by collinearities. However, the best models of Revised IP-OLDF and logistic regression that use PCA data are worse than the best models that use the original data.
2. Instead, we see no differences between the original and PCA data. Moreover, PCA data seems to obtain the best high-dimensional model.

## 3.4.2 Third Alternative Approach

We attempted to delete three variables related to collinearity. Table 3.12 lists the three eigenvectors from the 17th to the 19th component. These three components explain only 0.75 % of the total variance. Therefore, we consider Eqs. (3.8), (3.9), and (3.10). There are three variables groups from these equations, such as ($X12$, $X13$, $X14$), ($X4$, $X5$, $X7$, $X8$, $X9$), and ($X4$, $X5$, $X8$, $X9$), respectively. $X14$ is first excluded from ($X12$, $X13$, $X14$) by a stepwise technique from 19 to 1 variable. $X8$ and $X4$ are excluded from ($X4$, $X5$, $X7$, $X8$, $X9$) and ($X4$, $X5$, $X8$, $X9$), respectively. VIF of the 16-variable model is less than two. When we use SAS, we delete ($X4$, $X7$, $X14$) from the full model.

$$C19: 0.643X12 - 0.639X13 + 0.422X14 = 0 \qquad (3.8)$$

$$C18: 0.413X4 + 0.156X5 - 0.837X7 + 0.134X8 + 0.289X9 = 0 \qquad (3.9)$$

$$C17: -0.583X4 - 0.262X5 + 0.303X8 + 0.698X9 = 0 \qquad (3.10)$$

**Table 3.12** Eigenvectors of three components

| Var. | C17 | C18 | C19 |
|---|---|---|---|
| X1  | 0.004  | 0.001  | 0.001 |
| X2  | 0.006  | −0.007 | −0.003 |
| X3  | 0.005  | −0.002 | 0.001 |
| X4  | **−0.583** | **0.413** | 0.000 |
| X5  | **−0.262** | **0.156** | −0.004 |
| X6  | 0.050  | −0.048 | 0.000 |
| X7  | −0.049 | **−0.837** | 0.002 |
| X8  | **0.303** | **0.134** | 0.000 |
| X9  | **0.698** | **0.289** | 0.001 |
| X10 | 0.062  | 0.020  | 0.002 |
| X11 | 0.039  | −0.012 | −0.001 |
| X12 | 0.011  | 0.006  | **0.643** |
| X13 | −0.002 | 0.002  | **−0.639** |
| X14 | −0.022 | −0.002 | **0.422** |
| X15 | −0.017 | 0.000  | −0.001 |
| X16 | 0.019  | 0.012  | 0.001 |
| X17 | −0.015 | 0.006  | 0.001 |
| X18 | 0.023  | −0.003 | 0.000 |
| X19 | −0.010 | −0.005 | 0.000 |

**Table 3.13** MNM and seven "Diff1s"

| P | RIP | SVM4 | SVM1 | LP | IPLP | Logistic | LDF | QDF | RDA |
|---|---|---|---|---|---|---|---|---|---|
| 1 | 20 | 0 | 0 | 0 | 0 | **6** | 3 | 2 | 2 |
| 2 | 13 | 4 | 4 | 4 | 0 | **16** | 4 | 7 | 7 |
| 3 | 12 | 6 | 6 | 6 | 0 | **12** | 7 | 10 | 9 |
| 4 | 10 | 3 | 3 | 3 | 0 | **8** | 7 | **8** | **8** |
| 5 | 7 | **6** | 5 | 6 | 0 | 9 | **10** | 9 | 8 |
| 6 | 7 | **7** | 5 | 7 | 0 | 7 | **10** | 8 | 5 |
| 7 | 6 | 6 | 6 | 6 | 0 | 9 | **10** | 3 | 7 |
| 8 | 4 | 5 | 7 | 5 | 0 | 3 | **10** | 5 | 8 |
| 9 | 4 | 4 | 7 | 4 | 0 | 4 | **11** | 5 | 8 |
| 10 | 4 | 5 | 4 | 5 | 0 | 4 | **9** | 8 | 9 |
| 11 | 3 | **6** | 5 | 6 | 0 | 5 | **11** | 10 | 10 |
| 12 | 3 | 3 | 6 | 3 | 0 | 4 | **10** | 9 | 10 |
| 13 | 3 | 3 | 5 | 3 | 0 | 5 | **12** | 8 | 11 |
| 14 | 2 | 3 | 6 | 3 | 0 | 3 | **13** | 7 | 12 |
| 15 | 2 | 3 | 6 | 3 | 0 | 8 | **13** | 4 | 10 |
| 16 | 2 | 3 | 6 | 3 | 0 | 3 | **13** | 6 | 10 |

Table 3.13 lists MNM (RIP), and eight "Diff1s," which are the difference defined as (NMs of seven discriminant functions – MNM). Because NM of Revised IPLP-OLDF is equal to MNM, we omit its "Diff1" from the table. We can confirm that MNM decreases monotonously. Although NM of Revised IPLP-OLDF decreases monotonously, this result is not guaranteed theoretically. From the table, we find the following remarkable facts:

1. If we focus on the bold figures that are maximum Diff1, four models of logistic regression from one to four variables are worst and 12 models of Fisher's LDF from five to 16 variables are worst. With the exception of CPD data, logistic regression is better than Fisher's LDF, QDF, and RDA.
2. With the exception of CPD data, SVM4 is better than SVM1. However, SVM1 is better than SVM4 for five, six, ten, and eleven-variable models.
3. All NMs of Revised IPLP-OLDF are similar to MNMs.

Table 3.14 lists the results of seven LDFs by the Method 1. Six LDFs select the full models as the best models. M2s of RIP, LP, IPLP and LDF are minimum value 3.55 %. M2s of SVM4 and SVM1 are 3.80 and 4.57 %. Only logistic regression selects the two-variable model and its MNM is 3.95 %. Seven minus M1Diffs of logistic regression indicate the effects of Problem 1.

## 3.4 Trial to Remove Collinearity

**Table 3.14** Best Models by 100-fold cross-validation

| RIP | $M1$ | $M2$ | Diff. | F/B | |
|---|---|---|---|---|---|
| 1 | 7.92 | 9.04 | 1.12 | FB | |
| 2 | 4.83 | 6.03 | 1.21 | FB | |
| 3 | 4.38 | 5.95 | 1.57 | FB | |
| 4 | 3.58 | 5.65 | 2.07 | FB | |
| 5 | 2.56 | 4.61 | 2.05 | F | |
| 6 | 2.16 | 4.52 | 2.36 | B | |
| 7 | 1.94 | 4.70 | 2.76 | F | |
| 8 | 1.05 | 3.71 | 2.66 | B | |
| 9 | 0.82 | 3.60 | 2.79 | F | |
| 10 | 0.71 | 3.83 | 3.12 | B | |
| 11 | 0.61 | 3.84 | 3.23 | B | |
| 12 | 0.52 | 3.83 | 3.30 | F | |
| 13 | 0.40 | 3.90 | 3.50 | F | |
| 14 | 0.18 | 3.62 | 3.44 | B | |
| 15 | 0.09 | 3.59 | 3.50 | F | |
| 16 | 0.06 | **3.55** | 3.49 | B | |
| SVM4 | $M1$ | $M2$ | Diff. | $M1$Diff. | $M2$Diff. |
| 1 | 6.77 | 7.27 | 4.04 | 2.388 | 1.32 |
| 2 | 5.43 | 6.18 | 4.12 | 1.850 | 0.53 |
| 3 | 4.33 | 5.27 | 4.17 | 1.775 | 0.66 |
| 4 | 4.15 | 5.50 | 4.16 | 1.983 | 0.98 |
| 5 | 3.97 | 5.37 | 4.20 | 2.025 | 0.67 |
| 6 | 2.62 | 4.57 | 4.13 | 1.563 | 0.85 |
| 7 | 2.25 | 4.48 | 4.28 | 1.438 | 0.88 |
| 8 | 2.08 | 4.42 | 4.38 | 1.363 | 0.59 |
| 9 | 1.83 | 4.39 | 4.30 | 1.213 | 0.55 |
| 10 | 1.53 | 4.28 | 4.44 | 1.008 | 0.45 |
| 11 | 1.23 | 4.26 | 4.42 | 0.833 | 0.37 |
| 12 | 0.14 | **3.80** | 5.70 | 0.083 | **0.25** |
| SVM1 | $M1$ | $M2$ | Diff. | $M1$Diff. | $M2$Diff. |
| 1 | 6.77 | 7.27 | 0.50 | 2.388 | 1.33 |
| 2 | 5.43 | 6.18 | 0.75 | 1.850 | 0.53 |
| 3 | 4.38 | 5.30 | 0.92 | 1.825 | 0.69 |
| 4 | 4.29 | 5.50 | 1.21 | 2.125 | 0.98 |
| 5 | 4.08 | 5.39 | 1.32 | 2.133 | 0.69 |
| 6 | 3.11 | 4.90 | 1.80 | 2.054 | 1.19 |

(continued)

**Table 3.14** (continued)

| | | | | | |
|---|---|---|---|---|---|
| 7 | 2.93 | 4.93 | 2.00 | 2.108 | 1.33 |
| 8 | 2.85 | 5.02 | 2.17 | 2.138 | 1.19 |
| 9 | 2.78 | 5.18 | 2.41 | 2.163 | 1.34 |
| 10 | 2.43 | 4.85 | 2.41 | 1.913 | 1.02 |
| 11 | 2.10 | 4.72 | 2.62 | 1.708 | 0.83 |
| 12 | 1.61 | 4.63 | 3.03 | 1.425 | 1.01 |
| 13 | 1.32 | **4.57** | 3.25 | 1.263 | **1.02** |
| LP | $M1$ | M2 | Diff. | $M$1Diff. | $M$2Diff. |
| 1 | 6.52 | 6.86 | 0.34 | 1.692 | 0.83 |
| 2 | 6.77 | 7.27 | 0.50 | 2.388 | 1.32 |
| 3 | 5.43 | 6.18 | 0.75 | 1.850 | 0.53 |
| 4 | 4.34 | 5.28 | 0.94 | 1.783 | 0.67 |
| 5 | 4.15 | 5.49 | 1.35 | 1.983 | 0.97 |
| 6 | 3.97 | 5.38 | 1.41 | 2.029 | 0.68 |
| 7 | 2.63 | 4.53 | 1.90 | 1.571 | 0.82 |
| 8 | 2.25 | 4.45 | 2.21 | 1.429 | 0.85 |
| 9 | 2.08 | 4.41 | 2.34 | 1.363 | 0.58 |
| 10 | 1.83 | 4.37 | 2.54 | 1.213 | 0.52 |
| 11 | 1.53 | 4.23 | 2.70 | 1.008 | 0.41 |
| 12 | 1.22 | 4.22 | 3.00 | 0.825 | 0.33 |
| 13 | 0.59 | 3.83 | 3.24 | 0.404 | 0.21 |
| 14 | 0.13 | **3.55** | 3.42 | 0.071 | **0.00** |
| IPLP | $M$1 | $M$2 | Diff. | $M$1Diff. | $M$2Diff. |
| 1 | 4.83 | 6.08 | 1.25 | 0.004 | 0.05 |
| 2 | 4.38 | 6.00 | 1.62 | 0.000 | 0.05 |
| 3 | 3.58 | 5.54 | 1.95 | 0.004 | −0.11 |
| 4 | 2.57 | 4.47 | 1.90 | 0.013 | −0.15 |
| 5 | 2.18 | 4.53 | 2.35 | 0.017 | 0.01 |
| 6 | 1.95 | 4.67 | 2.71 | 0.013 | −0.03 |
| 7 | 1.06 | 3.70 | 2.65 | 0.004 | −0.01 |
| 8 | 0.83 | 3.64 | 2.82 | 0.008 | 0.04 |
| 9 | 0.71 | 3.73 | 3.02 | 0.000 | −0.10 |
| 10 | 0.61 | 3.73 | 3.12 | 0.000 | −0.11 |
| 11 | 0.53 | 3.83 | 3.30 | 0.004 | 0.00 |
| 12 | 0.39 | 3.93 | 3.54 | **−0.004** | 0.04 |
| 13 | 0.18 | 3.57 | 3.39 | 0.000 | −0.05 |
| 14 | 0.06 | **3.55** | 3.49 | 0.000 | **0.00** |

## 3.4 Trial to Remove Collinearity

**Table 3.14** (continued)

| Logistic | M1 | M2 | Diff. | M1Diff. | M2Diff. |
|---|---|---|---|---|---|
| 1 | 0.09 | 4.04 | 3.95 | −7.829 | −5.00 |
| 2 | 0.17 | **3.95** | 3.79 | **−4.658** | **-2.08** |
| 3 | 0.58 | 4.20 | 3.62 | **−3.804** | −1.75 |
| 4 | 0.78 | 4.25 | 3.47 | **−2.796** | −1.39 |
| 5 | 1.14 | 4.38 | 3.24 | **−1.421** | −0.24 |
| 6 | 1.00 | 4.40 | 3.40 | **−1.167** | −0.12 |
| 7 | 1.85 | 4.64 | 2.79 | **−0.092** | −0.06 |
| 8 | 1.33 | 4.52 | 3.19 | 0.275 | 0.80 |
| 9 | 2.04 | 4.73 | 2.69 | 1.225 | 1.13 |
| 10 | 2.03 | 4.73 | 2.70 | 1.321 | 0.90 |
| 11 | 2.24 | 4.74 | 2.51 | 1.625 | 0.90 |
| 12 | 2.53 | 4.88 | 2.35 | 2.013 | 1.06 |
| 13 | 2.53 | 4.73 | 2.20 | 2.138 | 0.84 |
| 14 | 3.93 | 5.89 | 1.96 | 3.750 | 2.27 |
| 15 | 2.73 | 4.78 | 2.04 | 2.646 | 1.19 |
| 16 | 4.23 | **5.88** | 1.65 | 4.167 | **2.33** |
| LDF | M1 | M2 | Diff. | M1Diff. | M2Diff. |
| 1 | 25.00 | 25.00 | **0.00** | 17.079 | 15.96 |
| 2 | 25.00 | 25.00 | **0.00** | 20.175 | 18.97 |
| 3 | 16.10 | 16.48 | 0.38 | 11.717 | 10.53 |
| 4 | 16.15 | 16.74 | 0.58 | 12.575 | 11.09 |
| 5 | 13.16 | 14.53 | 1.37 | 10.604 | 9.92 |
| 6 | 5.71 | 7.53 | 1.83 | 3.546 | 3.01 |
| 7 | 5.50 | 7.58 | 2.08 | 3.554 | 2.88 |
| 8 | 5.29 | 7.58 | 2.29 | 4.233 | 3.87 |
| 9 | 4.90 | 7.28 | 2.39 | 4.079 | 3.68 |
| 10 | 4.54 | 7.25 | 2.71 | 3.829 | 3.42 |
| 11 | 4.29 | 7.13 | 2.84 | 3.675 | 3.28 |
| 12 | 2.20 | 4.89 | 2.69 | 1.679 | 1.06 |
| 13 | 1.81 | 4.93 | 3.13 | 1.413 | 1.04 |
| 14 | 1.43 | 4.86 | 3.43 | 1.242 | 1.24 |
| 15 | 1.17 | 4.80 | 3.63 | 1.079 | 1.21 |
| 16 | 0.13 | **3.55** | 3.42 | 0.071 | **0.0** |

## 3.5 Summary

In this chapter, we discriminated CPD data with several difficult problems and solved these issues as follows:

1. There are strong collinearities in these data. We specified three collinearities in the data and established how to remove such collinearities.
2. We found a strange trend of NMs by QDF and found that QDF is fragile for collinearities. Moreover, NM of Fisher's LDF did not decrease in the 19 models from the one to 19-variable models selected by the forward stepwise procedure. On the other hand, NMs of our three MP-based OLDFs almost decreased.
3. In the original data, we selected the four-variable model as useful for the regression model selection procedure. However, the Method 1 recommends the nine-variable model as the best model. We think that many variables and/or collinearities cause this difference. Because the Iris data have four variables and might satisfy Fisher's assumption, the model selection procedure and best model selected the full model of eight LDFs. This is the reason we should no longer use the Iris data for evaluation.
4. We proposed a method for treating CPD data with collinearities by the best models. We compare three alternatives in Table 3.15. The basic Alternative 1 is the result of the original data from Table 3.7. The alternative 2 is to use the data modified by PCA in Table 3.11. The alternative 3 is to delete variables for the purpose of generating data without collinearities in Table 3.14. Table 3.15 indicates that the best model of Revised IP-OLDF by Alternative 3 is the best alternative. In near future, we must examine this result by other data.

In this chapter, we solved the previous problems. However, we found a new problem: CPD data has many OCPs. This fact implies that Revised IP-OLDF might search for several OCPs with the same MNMs and different coefficients groups that belong to different OCPs. This result means that it is difficult for us to evaluate the 95 % CI of discriminant coefficients. This problem is the sixth problem.

Table 3.15 Comparison of three data, such as original, PCA transformation, and 16 variables without collinearity

|  | Alternative 1 (original) |  | Alternative 2 (PCA) |  | Alternative 3 (delete collinearities) |  |
|---|---|---|---|---|---|---|
|  | Var. | M2 | Var. | M2 | Var. | M2 |
| Revised IP-OLDF | 9 | 3.64 | 19 | 4.03 | *16* | *3.55* |
| Logistic regression | 18 | 3.95 | 19 | 4.02 | 2 | 3.95 |

# References

Anderson E (1945) The irises of the Gaspe Peninsula. Bull Am Iris Soc 59:2–5
Cox DR (1958) The regression analysis of binary sequences (with discussion). J Roy Stat Soc B 20:215–242
Firth D (1993) Bias reduction of maximum likelihood estimates. Biometrika 80:27–39
Fisher RA (1936) The use of multiple measurements in taxonomic problems. Ann Eugenics 7:179–188
Fisher RA (1956) Statistical methods and statistical inference. Hafner Publishing Co, New Zealand
Flury B, Rieduyl H (1988) Multivariate statistics: a practical approach. Cambridge University Press, Cambridge
Friedman JH (1989) Regularized discriminant analysis. J Am Stat Assoc 84(405):165–175
Goodnight JH (1978) SAS technical report—the sweep operator: its importance in statistical computing—(R100). SAS Institute Inc, USA
Lachenbruch PA, Mickey MR (1968) Estimation of error rates in discriminant analysis. Technometrics 10:1–11
Miyake A, Shinmura S (1976) Error rate of linear discriminant function. In: Dombal FT, Gremy F (ed). North-Holland Publishing Company, The Netherland, pp 435–445
Miyake A, Shinmura S (1979) An algorithm for the optimal linear discriminant functions. Proceedings of the international conference on cybernetics and society, pp 1447–1450
Miyake A, Shinmura S (1980) An algorithm for the optimal linear discriminant function and its application. Jpn Soc Med Electron Biol Eng 18(6):452–454
Nomura Y, Shinmura S (1978) Computer-assisted prognosis of acute myocardial infarction. MEDINFO 77, In: Shires, W (ed), IFIP. North-Holland Publishing Company, The Netherland, pp 517–521
Sall JP (1981) SAS regression applications. SAS Institute Inc., USA. (Shinmura S. translate Japanese version)
Sall JP, Creighton L, Lehman A (2004) JMP start statistics, third edition. SAS Institute Inc., USA. (Shinmura S. edits Japanese version)
Schrage L (1991) LINDO—an optimization modeling systems. The Scientific Press, USA. (Shinmura S. & Takamori, H. translate Japanese version)
Schrage L (2006) Optimization modeling with LINGO. LINDO Systems Inc. USA. (Shinmura S. translates Japanese version)
Shimizu T, Tsunetoshi Y, Kono H, Shinmura S (1975) Classification of subjective symptoms of junior high school Students affected by photochemical air pollution. J Jpn Soc Atmos Environ, 9/4:734–741. Translated for NERCLibrary, EPA, from the original Japanese by LEO CANCER Associates, P.O.Box 5187 Redwood City, California 94063, (TR 76-213)
Shinmura S, Miyake A (1979) Optimal linear discriminant functions and their application. COMPSAC 79:167–172
Shinmura S, Suzuki T, Koyama H, Nakanishi K (1983) Standardization of medical data analysis using various discriminant methods on a theme of breast diseases. MEDINFO 83, In: Vann Bemmel JH, Ball MJ, Wigertz O (ed). North-Holland Publishing Company, The Netherland, pp 349–352
Shinmura S, Iida K, Maruyama C (1987) Estimation of the effectiveness of cancer treatment by SSM using a null hypothesis model. Inf Health Social Care 7(3):263–275. doi:10.3109/14639238709010089
Shinmura S (1998) Optimal linear discriminant functions using mathematical programming. J Jpn Soc Comput Stat 11/2:89–101
Shinmura S, Tarumi T (2000) Evaluation of the optimal linear discriminant functions using integer programming (IP-OLDF) for the normal random data. J Jpn Soc Comput Stat 12(2):107–123
Shinmura S (2000a) A new algorithm of the linear discriminant function using integer programming. New Trends Prob Stat 5:133–142

Shinmura S (2000b) Optimal linear discriminant function using mathematical programming. Dissertation, March 200:1–101, Okayama University, Japan

Shinmura S (2001) Analysis of effect of SSM on 152,989 cancer patient. ISI2001: 1–2. doi:10.13140/RG.2.1.3077.9281

Shinmura S (2003) Enhanced algorithm of IP-OLDF. ISI2003 CD-ROM, pp 428–429

Shinmura S (2004) New algorithm of discriminant analysis using integer programming. IPSI 2004 Pescara VIP Conference CD-ROM, pp 1–18

Shinmura S (2005) New age of discriminant analysis by IP-OLDF—beyond Fisher's linear discriminant function. ISI2005, pp 1–2

Shinmura S (2007) Overviews of discriminant function by mathematical programming. J Jpn Soc Comput Stat 20(1–2):59–94

Shinmura S (2010a) The optimal linearly discriminant function (Saiteki Senkei Hanbetu Kansuu). Union of Japanese Scientist and Engineer Publishing, Japan

Shinmura S (2010b) Improvement of CPU time of revised IP-OLDF using linear programming. J Jpn Soc Comput Stat 22(1):39–57

Shinmura S (2011a) Beyond Fisher's linear discriminant analysis—new world of the discriminant analysis. ISI2011 CD-ROM, pp 1–6

Shinmura S (2011b) Problems of discriminant analysis by mark sense test data. Jpn Soc Appl Stat 40(3):157–172

Shinmura S (2013) Evaluation of optimal linear discriminant function by 100-fold cross-validation. ISI CD-ROM, pp 1–6

Shinmura S (2014a) End of discriminant functions based on variance-covariance matrices. ICORE2014, pp 5–16

Shinmura S (2014b) Improvement of CPU time of linear discriminant functions based on MNM criterion by IP. Stat Optim Inf Comput 2:114–129

Shinmura S (2014c) Comparison of linear discriminant functions by $K$-fold cross-validation. Data Anal 2014:1–6

Shinmura S (2015a) The 95 % confidence intervals of error rates and discriminant coefficients. Stat Optim Inform Comput 2:66–78

Shinmura S (2015b) A trivial linear discriminant function. Stat Optim Inf Comput 3:322–335. doi:10.19139/soic.20151202

Shinmura S (2015c) Four serious problems and new facts of the discriminant analysis. In: Pinson E, Valente F, Vitoriano B (ed) Operations research and enterprise systems: 15–30. Springer, Berlin. (ISSN: 1865-0929, ISBN: 978-3-319-17508-9, doi:10.1007/978-3-319-17509-6)

Shinmura S (2015d) Four problems of the discriminant analysis. ISI 2015:1–6

Shinmura S (2016) The best model of Swiss banknote data. Stat Optim Inf Comput 4:118–131. International Academic Press (ISSN: 2310-5070 (online) ISSN: 2311-004X (print), doi:10.19139/soic.v4i2.178)

VapnikV (1995) The nature of statistical learning theory. Springer, Berlin

# Chapter 4
# Student Data and Problem 1

## 4.1 Introduction

Student data (Shinmura 2010a) consist of 40 students with six variables, which are the study hours per day ($X1$), spending money per month ($X2$), drinking days per week ($X3$), gender ($X4$), smoking ($X5$), and examination scores ($X6$). The amount of data is not large. We published four statistical books on SAS, SPSS, Statistica, and JMP using these data because the reader could easily understand the meaning of variables and data. Although we never believed that these data would be helpful for our research, we discriminated the data after completing analysis of the Iris (Anderson 1945), cephalo-pelvic disproportion (CPD) (Miyake and Shinmura 1980), and random number data (Shinmura and Tarumi 2000) in 1999. When we discriminated these data using five variables with 70 points as the passing mark (Score $\geq$ 70), we found a defect in optimal linear discriminant function using integer programming (IP-OLDF). Because four numerical variables are integer values and two variables are the binary integers 0/1, and there are many overlapping cases, the obtained vertex of convex polyhedron (CP) consists of over ($p + 1$) cases and the solution is not true minimum number of misclassifications (MNMs). Moreover, three optimal CPs (OCPs) were found by the K-option of LINGO (Schrage 1991, 2006), as shown in Fig. 4.3. Although we recognized Problems 1 and 4 before 1980, we did not realize that Problem 1 causes a defect in IP-OLDF (Shinmura 1998, 2000a, b, 2003, 2004, 2005). By the scatter plot of the two variables $X1$ and $X2$ indicated in Table 1.1, we found that the reason for the defect in IP-OLDF is the result of Problem 1. However, we could not revise it until 2006, when Revised IP-OLDF (Shinmura 2007, 2010a) solved Problem 1 completely. In 2004, IP-OLDF found that Swiss banknote data (Flury and Rieduyl 1988) are linearly separable data (LSD), and no LDFs, with the exception of Revised IP-OLDF and hard-margin support vector machine (H-SVM), could discriminate LSD theoretically (Problem 2). In 2005, we were able to validate the discrimination of

---

Whole my studies and books are listed in "http://researchmap.jp/read0049917/."

original data (the training sample) by 20,000 resampling samples (the validation sample). After 2006, we could compare six MP-based LDFs and two statistical LDFs. The six MP-based LDFs are Revised IP-OLDF, Revised LP-OLDF, Revised IPLP-OLDF (Shinmura 2010b, 2014b), and three SVMs (Vapniik 1995), and the two statistical LDFs (Sall et al. 2004) are Fisher's LDF (Fisher 1936, 1956) and logistic regression (Cox 1958; Firth 1993). Although quadratic discriminant function (QDF) and a regularized discriminant analysis (RDA) (Friedman 1989) are not LDFs, these discriminant functions discriminate Student data. After 2009, we developed the 100-fold cross-validation for small sample method (the Method 1). The Method 1 solved Problem 4 (Shinmura 2010a, 2013, 2014c, 2015a, b), and the best model provided a clear evaluation of eight LDFs (Shinmura 2016a, c). Although we could not explain the useful meaning of 95 % CI of the coefficient, we completed the basic research in 2010 (Shinmura 2010a). After 2010, applied research started on LSD discrimination using pass/fail determination (Shinmura 2011) and Japanese-automobile data. In 2010, we found that the pass/fail determination using examination scores caused Problem 3, and we solved it in 2013. The defect of the generalized inverse matrices caused Problem 3 for QDF. In 2015, the applied research was completed because we successfully explained the useful meaning of 95 % CI of the coefficient, and the Method 1 solved Problem 4 completely (Shinmura 2014a, 2015c, d). In Oct. 2015, young researcher, Ishii et al. (2014), presented the challenging results of microarray datasets using PCA. Because Jeffery et al. (2006) indicated six microarray datasets (the datasets) on HP,[1] we developed the Matroska feature-selection method (the Method 2) within 41 days (Shinmura 2016b). For more than ten years, many researchers have struggled with the analysis of the dataset because the dataset consists of few cases with huge genes ($n \ll p$) (Problem 5). The Theory is most suitable for gene analysis. Recently, many researchers have expected LASSO (Simon et al. 2013) to solve Problem 5. Because Revised IP-OLDF selects gene features naturally, LASSO researchers should compare their results to ours through the Swiss banknote data, Japanese-automobile data, Student linearly separable data, and six microarray datasets. Such comparison should be helpful for LASSO research. The platform for this book is over 12 different types of datasets. Eight LDFs play on this platform, which is a different scenario from statistical themes.

## 4.2 Student Data

### 4.2.1 Data Outlook

Table 4.1 lists Student data that consist of two classes: 25 students who passed the examination and 15 students who failed, where the passing mark is 70 points. There

---

[1] http://www.bioinf.ucd.ie/people/ian/.

## 4.2 Student Data

**Table 4.1** Student data (Eight bold cases are on the discriminant hyperplane)

| SN | X1 | X2 | X3 | C  | Y (score) | X4 (gender) | X5 (smoke) |
|----|----|----|----|----|-----------|-------------|------------|
| 1  | 10 | 2  | 0  | 1  | 90  | 1 | 0 |
| 2  | 9  | 2  | 0  | 1  | 100 | 1 | 0 |
| 3  | 5  | 2  | 1  | 1  | 75  | 0 | 1 |
| 4  | 7  | 3  | 1  | 1  | 70  | 1 | 1 |
| 5  | 3  | 3  | 1  | 1  | 85  | 1 | 0 |
| 6  | 7  | 3  | 0  | 1  | 90  | 0 | 1 |
| 7  | 7  | 3  | 0  | 1  | 90  | 0 | 0 |
| 8  | 7  | 3  | 0  | 1  | 95  | 1 | 0 |
| 9  | 6  | 3  | 2  | 1  | 80  | 0 | 1 |
| 10 | 3  | 3  | 3  | 1  | 70  | 0 | 1 |
| 11 | 6  | 3  | 0  | 1  | 85  | 1 | 1 |
| 12 | 6  | 3  | 2  | 1  | 70  | 1 | 0 |
| 13 | 8  | 3  | 0  | 1  | 85  | 1 | 0 |
| 14 | 5  | 3  | 2  | 1  | 75  | 1 | 0 |
| 15 | 2  | 4  | 2  | 1  | 80  | 1 | 0 |
| 16 | 5  | 4  | 4  | 1  | 75  | 0 | 1 |
| 17 | 12 | 4  | 1  | 1  | 100 | 0 | 0 |
| 18 | 4  | 4  | 1  | 1  | 70  | 1 | 0 |
| 19 | 10 | 4  | 3  | 1  | 80  | 0 | 0 |
| 20 | 7  | 4  | 1  | 1  | 75  | 1 | 0 |
| 21 | 5  | 4  | 1  | 1  | 85  | 0 | 0 |
| 22 | *6*  | *5*  | 1  | 1  | 85  | 0 | 0 |
| 23 | *9*  | *5*  | 1  | 1  | 75  | 0 | 0 |
| 24 | *4*  | *5*  | 1  | 1  | 70  | 1 | 1 |
| 25 | *3*  | *5*  | 1  | 1  | 75  | 0 | 1 |
| 26 | 3  | 2  | 1  | -1 | 60  | 0 | 1 |
| 27 | 5  | 2  | 1  | -1 | 60  | 1 | 0 |
| 28 | 3  | 3  | 2  | -1 | 65  | 0 | 1 |
| 29 | *5*  | *5*  | 3  | -1 | 65  | 1 | 0 |
| 30 | *2*  | *5*  | 4  | -1 | 60  | 0 | 1 |
| 31 | *3*  | *5*  | 4  | -1 | 65  | 1 | 1 |
| 32 | *2*  | *5*  | 4  | -1 | 40  | 0 | 1 |
| 33 | 2  | 6  | 3  | -1 | 55  | 0 | 0 |
| 34 | 1  | 6  | 5  | -1 | 60  | 0 | 1 |
| 35 | 4  | 6  | 2  | -1 | 65  | 1 | 1 |
| 36 | 3  | 6  | 2  | -1 | 60  | 0 | 0 |
| 37 | 3  | 7  | 5  | -1 | 55  | 0 | 1 |
| 38 | 3  | 7  | 3  | -1 | 50  | 1 | 1 |
| 39 | 1  | 8  | 7  | -1 | 60  | 0 | 1 |
| 40 | 3  | 10 | 6  | -1 | 40  | 0 | 1 |

are five variables: X1 is the number of hours of study per day (hours/day); X2 is the amount of money spent per month (10,000 yen/month); X3 are some days drinking per week (day/week); X4 is gender; and X5 is smoking. The passing score is 70 points. Gender (male = 1, female = 0) and smoke (smoking = 1, non-smoking = 0) are dummy variables. The amount of data is not large. First, there is resistance against using this data in the research. However, we found a serious defect in IP-OLDF (Problem 1) that stops optimization by setting over $p$-cases on the discriminant hyperplane in the case of $p$-variable model. S-SVM, Revised LP-OLDF, and Revised IPLP-OLDF stop optimization by setting cases on two SVs. These MP-based LDFs cannot theoretically avoid some cases on the discriminant hyperplane.

### 4.2.2 Different LDFs

We discriminated the two-variable model (X1, X2) by IP-OLDF in 2006. What's Best!, an Excel add-in solver, formulated IP-OLDF. As shown in Fig. 4.1, IP-OLDF selected $X2 = 5$ as the discriminant hyperplane, which appears as a straight line, and outputted MNM = 3 in Fig. 4.1. Four passing students and four failing students spent 50,000 yen/month, as indicated in Table 4.1. These eight students are on the discriminant hyperplane of IP-OLDF and are treated as a classified group because eight $e_i$ in Eq. (1.9) are zeroes. This fact means that IP-OLDF classifies four passing students into the passing class, and four failing students into the failing class, which is obviously nonsense. To this point, we have not been able to discriminate students on the discriminant hyperplane exactly if there are over three ($p + 1$) students on the discriminant hyperplane. On the other hand, Revised IP-OLDF finds three true OCPs, genuine MNMs (Miyake and

**Fig. 4.1** Defect in IP-OLDF caused by Problem 1 (IP-OLDF: *straight horizontal line*, Revised IP-OLDF: *straight line with slope*, Fisher's LDF: *dashed line*, Revised LP-OLDF: *dotted line*) (Shinmura 2010a)

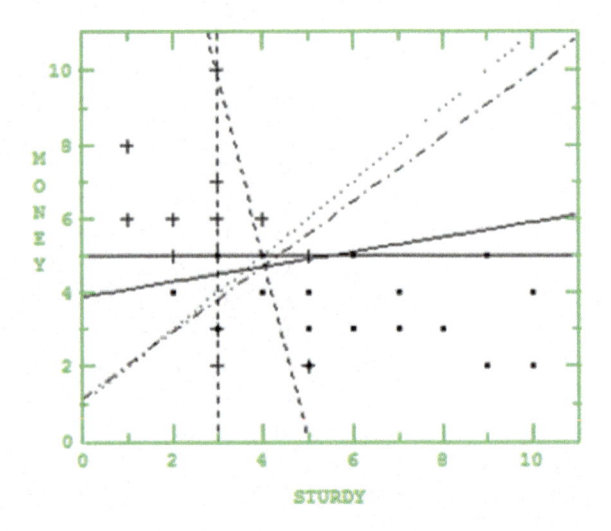

## 4.2 Student Data

Shinmura 1976, 1979; Shinmura and Miyake 1979) of which are five by LINGO k-best option in Fig. 4.3 (Schrage 2006). One of the three Revised IP-OLDFs is a straight line with slope. Fisher's LDF is a dashed line, and Revised LP-OLDF is a dotted line.

Figure 4.2 is a contour plot of the two classes that correspond to Fig. 4.1. The two ellipses include 95 % of each type of student. The passing class (symbol: ".") is spread on the $X1$ axis and the failing class (symbol: "×") is spread on the $X2$ axis. The two classes are substantially perpendicular to each other.

### 4.2.3 Comparison of Seven LDFs

If we analyze the two-variable model ($X1$, $X2$) by IP-OLDF, it would select $X2 = 5$ as the discriminant hyperplane and output MNM = 3. IP-OLDF searches for the vertex of true OCP when data are general positions and there are $p=2$ students on the discriminant hyperplane. However, IP-OLDF searched for the vertex composed of eight students. If it searches for the vertex that consists of two students, it would find that only four CPs shared this vertex. Each CP interior point is located in the four plus/minus side of the linear hyperplane $H_i(\mathbf{b}) = y_i \times (\mathbf{x}_i\mathbf{b} + 1) = 0$, as shown in Eq. (1.9). The CP interior points are ($X1$, $X2$) = (+, +), (+, −), (−, +), and (−, −). If we select the CP with (+, +), this is the true OCP. In this case, because 16 (= 2 × 8) CPs share this vertex composed of eight linear hyperplanes $H_i(\mathbf{b})$, the solution might not be the correct vertex of OCP. Revised IP-OLDF finds true OCP, true MNM of which is five. If we use LINGO k-best option, as shown in Fig. 4.3, we know that there are three OCPs. In Chap. 3, we introduced several OCPs that cause deep understanding of the 95 % CI discriminant coefficient (Problem 6).

**Fig. 4.2** Two 95 % ellipses (passing and failing classes are spread on $X1$ axis and $X2$ axis, respectively)

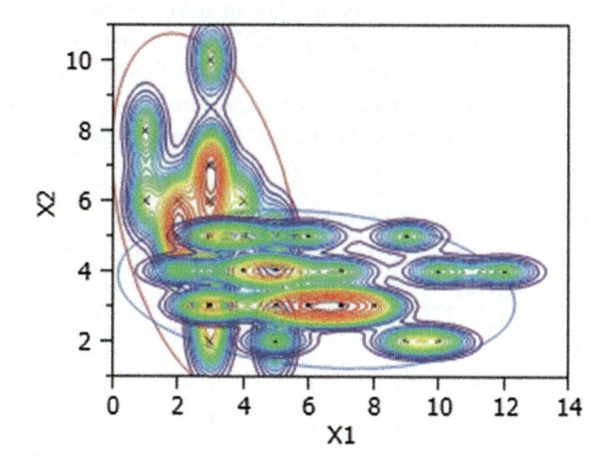

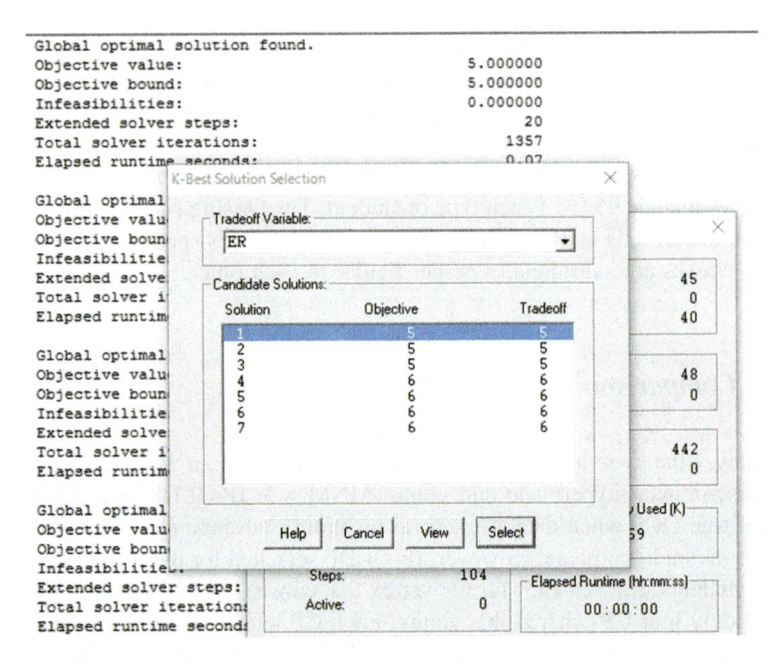

**Fig. 4.3** *K*-best solutions

Table 4.2 lists the comparison of MNM and eight "Diff1s" instead of eight NMs. "Diff1" is the difference defined as (NMs of eight functions − MNM). Revised IPLP-OLDF is equal to MNM. The two SVM1 models are worse than those of SVM4. The two NMs of Revised LP-OLDF are less than MNM because two differences are negative values, as is evident in the column "$f = 0$." Although Problem 1 causes this result, we cannot check the four statistical discriminant functions because statistical developers are unfamiliar with Problem 1. Even if "Diff1s" are greater than or equal to zero, we cannot find Problem 1. Therefore, we output the number of students on the discriminant hyperplane such as column "$f=0$" of all discriminant functions, except for Revised IP-OLDF and H-SVM. The four models of Revised LP-OLDF have positive values, and they are not free from Problem 1. Although three more MP-based LDFs are free from Problem 1 in this

**Table 4.2** Comparison between MNM and "Diff1s"

| SN | Var. | RIP | SVM4 | SVM1 | LP | $f = 0$ | IPLP | Logistic | LDF | QDF | RDA |
|---|---|---|---|---|---|---|---|---|---|---|---|
| 1 | 1–3 | 3 | 1 | 1 | 1 | 0 | 0 | 3 | 3 | 1 | 2 |
| 2 | 2–3 | 3 | 1 | 2 | 1 | 3 | 0 | 1 | 3 | 2 | 2 |
| 3 | 1, 3 | 5 | 1 | 2 | 3 | 0 | 0 | 0 | 0 | 2 | 1 |
| 4 | 1, 2 | 5 | 1 | 1 | 1 | 1 | 0 | 2 | 1 | 2 | 2 |
| 5 | 1 | 7 | 1 | 1 | −1 | 3 | 0 | 0 | 1 | 1 | 1 |
| 6 | 2 | 7 | 0 | 0 | −4 | 8 | 0 | 0 | 0 | 0 | 0 |
| 7 | 3 | 8 | 0 | 0 | 0 | 0 | 0 | 1 | 0 | 0 | 0 |

## 4.2 Student Data

data, we cannot determine the four statistical discriminant functions. If the number of cases ("$h$") on the discriminant hyperplane is greater than or equal to one, we should consider that true NM might increase, i.e., "current NM + $h$."

### 4.2.4 K-Best Option

In the IP model, such as Revised IP-OLDF, we should determine whether there are several global solutions by "$K$-best option," as shown in Fig. 4.3. After we select "LINGO option $\rightarrow$ Integer Solver $\rightarrow$ K-Best Solution," we set "$K$-Best = 7." The LINGO output is shown in Fig. 4.3. There are three global optimum models with "object value = 5 for two-variable model (X1, X2)." If we select the first solution, we obtain the first discriminant hyperplane, such as $X2 = X/2 + 4$. The second solution is "$X2 = X1/4 + 3.63$," and the third solution is "$X2 = X/2 + 3.25$."

When we check seven models, we know that two of the models have several global minimum solutions, as indicated in Table 4.3. The sixth one-variable model ($X2$) has two OCPs with MNMs of seven.

### 4.2.5 Evaluation by Regression Analysis

Before 2005, we attempted to evaluate several discriminant functions by the original data as training samples because we had no validation samples. Table 4.4 lists the results of eight regression analyses (Sall 1981), which were some of the validation approaches used. MNMs regressed eight NMs. Because the NMs of Revise IPLP-OLDF are equal to MNMs, the regression line is "IPLP = $0 + 1 \times$ RIP" and $R$-square is equal to one. If we evaluate the eight discriminant functions at MNM = 3, 5, and 7, the superiority or inferiority is drastically changed. For example, IPLP is best at MNM = 3. However, it is worst at MNM = 7, although MNM is minimum value among all NMs.

**Table 4.3** MNMs and OCPs of seven models

| SN | X1 | X2 | X3 | c | MNM | OCP |
|---|---|---|---|---|---|---|
| 1 | 1 | 1 | 1 | 1 | 4 | 1 |
| 2 | 0 | 1 | 1 | 1 | 3 | 1 |
| 3 | 1 | 0 | 1 | 1 | 5 | 1 |
| 4 | 1 | 1 | 0 | 1 | 5 | 3 |
| 5 | 1 | 0 | 0 | 1 | 7 | 1 |
| 6 | 0 | 1 | 0 | 1 | 7 | 2 |
| 7 | 0 | 0 | 1 | 1 | 8 | 1 |

We obtain various regression coefficients in the range of [0.39, 1]. Therefore, we must evaluate the superiority or inferiority of eight results, as shown in Fig. 4.4. There are at least five interactions at MNM = 4.5, 5.5, 5.8 7.5, and 8. From Fig. 4.4, the superiority or inferiority of eight results must be evaluated in at least six different segments. In the range of MNM < 4.5, we can evaluate intercepts, such as IPLP < SVM4 (SVM1) < LP < logistic < LP < QDF < RDA < LDF, because we can confirm by the predict value at MNM = 3 in Table 4.4.

**Table 4.4** Eight regression analyses

|  | $c$ | $b$ | $R^2$ | MNM = 3 | MNM = 5 | MNM = 7 |
|---|---|---|---|---|---|---|
| SVM4 | 1.88 | 0.76 | 0.976 | 4.16 | 5.68 | 7.59 |
| SVM1 | 1.88 | 0.76 | 0.976 | 4.16 | 5.68 | 7.59 |
| LP | 3.45 | 0.39 | 0.15 | 4.62 | 5.40 | 1.44 |
| IPLP | 0 | 1 | 1 | 3 | 5.00 | 7.00 |
| Logistic | 2.83 | 0.66 | 0.66 | 4.81 | 6.13 | 5.28 |
| LDF | 4.22 | 0.43 | 0.578 | 5.51 | 6.37 | 4.48 |
| QDF | 2.84 | 0.69 | 0.816 | 4.91 | 6.29 | 6.40 |
| RDA | 3.3 | 0.6 | 0.886 | 5.1 | 6.30 | 6.80 |

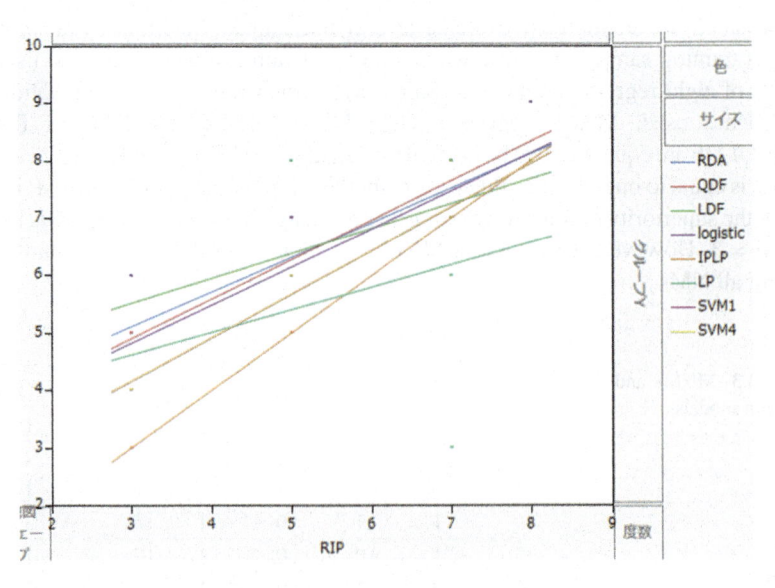

**Fig. 4.4** Comparison of eight discriminant functions by MNM

## 4.3 100-Fold Cross-Validation of Student Data

### 4.3.1 Best Model

We examine seven LDFs by the Method 1 in Table 4.5. The column "Var" is the suffix for variables. $M1$ decreases monotonously from one to three variables. There are six model sequences, such as (1) → (1, 2)/(1, 3) → (1, 2, 3), 2 → (1, 2)/(2, 3) → (1, 2, 3), 3 → (1, 3)/(2, 3) → (1, 2, 3). Three and one "M1Diffs" of Revised LP-OLDF and logistic regression are negative. This indicates that Revised LP-OLDF and logistic regression might not be free from Problem 1 because $M1$ of Revised IP-OLDF is the minimum value among all $M1$s, similar to MNM. Because $M2$ of the two-variable model ($X2$, $X3$) of Revised IP-OLDF is the minimum $M2$ among all $M2$s, we determine that this model is the best. We compare Revised IP-OLDF with six LDFs by this model; "$M2$Diff" of SVM4, SVM1, LP, IPLP, Logistic, and LDF is 3.55, 3.55, 2.55, −0.32, 8.52, and 10.53 %, respectively. These results, with the exception of Revised IPLP-OLDF, are poor. The results of logistic regression and Fisher's LDF are particularly poor. Because Revised IPLP-OLDF discriminates data by Revised LP-OLDF in the first step, Revised

**Table 4.5** Best model by 100-fold cross-validation for small sample (Bold figures show the seven best models)

| RIP | | M1 | M2 | Var. | |
|---|---|---|---|---|---|
| 1m21s | 1 | 5.90 | 12.00 | 1, 2, 3 | |
| | 2 | 9.25 | 14.93 | 1, 2 | |
| | 3 | 10.50 | 15.48 | 1, 3 | |
| | 4 | 16.40 | 18.90 | 1 | |
| | 5 | 7.40 | **9.25** | 2, 3 | |
| | 6 | 14.98 | 17.50 | 2 | |
| | 7 | 17.90 | 21.33 | 3 | |
| SVM4 | | M1 | M2 | M1Diff. | M2Diff. |
| 38s | 1 | 9.20 | 14.48 | 3.30 | 2.48 |
| | 2 | 13.38 | 16.68 | 4.13 | 1.75 |
| | 3 | 14.30 | 17.75 | 3.80 | 2.28 |
| | 4 | 17.60 | 18.50 | 1.20 | −0.40 |
| | 5 | 10.63 | **12.80** | 3.23 | **3.55** |
| | 6 | 15.88 | 17.90 | 0.90 | 0.40 |
| | 7 | 19.55 | 20.40 | 1.65 | −0.93 |
| SVM | | M1 | M2 | M1Diff. | M2Diff. |
| 138s | 1 | 9.20 | 14.48 | 3.30 | 2.48 |
| | 2 | 13.38 | 16.68 | 4.13 | 1.75 |
| | 3 | 14.30 | 17.75 | 3.80 | 2.28 |
| | 4 | 17.60 | 18.50 | 1.20 | −0.40 |
| | 5 | 10.63 | **12.80** | 3.23 | **3.55** |
| | 6 | 15.88 | 17.90 | 0.90 | 0.40 |
| | 7 | 19.55 | 20.40 | 1.65 | −0.93 |

(continued)

**Table 4.5** (continued)

| | | | M1 | M2 | M1Diff. | M2Diff. |
|---|---|---|---|---|---|---|
| LP 35s | | 1 | 8.98 | 13.85 | 3.08 | 1.85 |
| | | 2 | 12.88 | 15.85 | 3.63 | 0.92 |
| | | 3 | 13.28 | 16.45 | 2.78 | 0.98 |
| | | 4 | 15.73 | 16.40 | −0.67 | −2.50 |
| | | 5 | 9.95 | **11.80** | 2.55 | **2.55** |
| | | 6 | 11.43 | 13.00 | −3.55 | −4.50 |
| | | 7 | 16.63 | 17.48 | −1.28 | −3.85 |
| IPLP 1m23s | | | M1 | M2 | M1Diff. | M2Diff. |
| | | 1 | 5.95 | 11.68 | 0.05 | −0.32 |
| | | 2 | 9.38 | 15.28 | 0.13 | 0.35 |
| | | 3 | 10.75 | 15.53 | 0.25 | 0.05 |
| | | 4 | 16.40 | 19.05 | 0.00 | 0.15 |
| | | 5 | 7.40 | **8.93** | 0.00 | **−0.32** |
| | | 6 | 14.98 | 17.50 | 0.00 | 0.00 |
| | | 7 | 17.98 | 21.55 | 0.08 | 0.23 |
| logistic 5m | | | M1 | M2 | M1Diff. | M2Diff. |
| | | 1 | 11.23 | **15.41** | 5.33 | 3.41 |
| | | 2 | 14.23 | 16.63 | 4.98 | 1.71 |
| | | 3 | 15.10 | 16.87 | 4.60 | 1.40 |
| | | 4 | 12.18 | 14.68 | −4.23 | −4.22 |
| | | 5 | 17.55 | 17.77 | 10.15 | **8.52** |
| | | 6 | 15.50 | 17.45 | 0.53 | −0.05 |
| | | 7 | 19.60 | 20.08 | 1.70 | −1.25 |
| LDF 3m | | | M1 | M2 | M1Diff. | M2Diff. |
| | | 1 | 26.75 | 15.93 | 20.85 | 3.93 |
| | | 2 | 30.55 | 17.64 | 21.30 | 2.72 |
| | | 3 | 28.80 | **15.66** | 18.30 | 0.19 |
| | | 4 | 23.80 | 16.04 | 7.40 | −2.86 |
| | | 5 | 39.43 | 19.78 | 32.03 | **10.53** |
| | | 6 | 35.53 | 18.27 | 20.55 | 0.77 |
| | | 7 | 39.13 | 19.86 | 21.23 | −1.47 |

IPLP-OLDF is not free from Problem 1 as same as Revised LP-OLDF. Although Table 4.2 lists only four cases on the discriminant hyperplane of Revised LP-OLDF, we can not know the effect of the Method 1 because we do not check the numbers of f(x)=0 for the 100 training and validation samples.

## 4.3.2 Comparison of Coefficients by LINGO Program 1 and Program 2

Table 4.6 consists of three types of coefficients lists by LINGO Program 1, Program 2 and JMP script. First, LINGO Program 1 outputs coefficients of RIP, SVM4 and IPLP. Because SVM1 and LP are almost same as SVM4, these two LDFs are omit. Last column is the intercept. Second, LINGO Program 2 outputs the 95 % CI of coefficients by RIP 100 and SVM4/100, that include above original coefficients. Although we expect original coefficient are similar to the medians of 95 % CI, our expectation is no correct. Three one-variable models of Revised IP-OLDF (RIP100) and SVM4 (SVM4/100) are rejected at 5 % level. Third, JMP output Fisher's LDF and logistic regression show the coefficient and SE in the first and second rows. Although Fisher never formulated the SE equation for Fisher's LDF (Problem 4), we obtain SEs by the regression analysis through the plug-in rule1. The Hessian matrix obtains SEs for logistic regression. Three one-variable models of Fisher's LDF and logistic are rejected at 5 % level, also. The 95 % CI of "RIP100, SVM4/100, LDF and Logistic" suggest us only one-variable models are meaningful in practical use.

**Table 4.6** 100-fold cross-validation

|      | SN | X1      | X2     | X3     | C |
|------|----|---------|--------|--------|---|
| RIP  | 1  | −0.25   | −0.143 | −0.059 | 1 |
|      | 2  | 0.125   | −0.25  |        | 1 |
|      | 3  | 0.125   |        | −0.751 | 1 |
|      | 4  | −0.286  |        |        | 1 |
|      | 5  |         | −0.2   | −6E-5  | 1 |
|      | 6  |         | −0.167 |        | 1 |
|      | 7  |         |        | −0.4   | 1 |
| SVM4 | 1  | 0.077   | −0.154 | −0.231 | 1 |
|      | 2  | 1       | −1     |        | 1 |
|      | 3  | 1       |        | −2     | 1 |
|      | 4  | −0.25   |        |        | 1 |
|      | 5  |         | −0.1   | −0.2   | 1 |
|      | 6  |         | −0.2   |        | 1 |
|      | 7  |         |        | −0.4   | 1 |
| IPLP | 1  | 0       | −0.2   | −6E-5  | 1 |
|      | 2  | 0.0002  | −0.2   |        | 1 |
|      | 3  | 0.125   |        | −0.75  | 1 |
|      | 4  | −0.29   |        |        | 1 |
|      | 5  |         | −0.2   | −6E-5  | 1 |
|      | 6  |         | −0.18  | 0      | 1 |
|      | 7  |         |        | −0.38  | 1 |

(continued)

**Table 4.6** (continued)

|  | SN | X1 | X2 | X3 | C |
|---|---|---|---|---|---|
| RIP100 | 1 | −1.5/−0.02/0.65 | −0.43/−0.15/0.67 | −3.33/−0.06/2.4 | 1 |
|  | 2 | −0.62/0/176 | −119/−0.22/0.35 |  | 1 |
|  | 3 | −2/−0.32/2.61 |  | −0.694/0.11/4 | 1 |
|  | 4 | −0.4/**−0.29**/−0.18 |  |  | 1 |
|  | 5 |  | −0.38/−0.17/0.58 | −2.32/−0.06/0.09 | 1 |
|  | 6 |  | −0.25/**−0.2**/−0.17 |  | 1 |
|  | 7 |  |  | −0.82/**−0.4**/−0.22 | 1 |
| SVM4/100 | 1 | −0.99/0.03/3.48 | −1.48/−0.15/0.4 | −4.16/−0.1/1.39 | 1 |
|  | 2 | −1.05/0/2 | −2/−0.22/0.53 |  | 1 |
|  | 3 | −1E6/−0.25/8E7 |  | −3E8/0/3286607 | 1 |
|  | 4 | −0.29/**−0.25**/−0.22 |  |  | 1 |
|  | 5 |  | −0.24/−0.12/0 | −0.58/−0.15/0 | 1 |
|  | 6 |  | −0.25/**−0.2**/−0.15 |  | 1 |
|  | 7 |  |  | −0.5/−0.4/−0.27 | 1 |
| LDF | 1 | **0.134** | −0.095 | −0.163 | 0.315 |
|  |  | 0.05* | 0.1 | 0.11 | 0.49 |
|  | 2 | 0.169 | **−0.192** |  | 0.225 |
|  |  | 0.17 | 0.07* |  | 0.23 |
|  | 3 | **0.137** |  | **−0.233** | 0.042 |
|  |  | 0.05* |  | 0.08* | 0.41 |
|  | 4 | **0.23** |  |  | −0.894 |
|  |  | 0.05* |  |  | 0.26* |
|  | 5 |  | −0.106 | **−0.277** | 1.261 |
|  |  |  | 0.1 | 0.1* | 0.34* |
|  | 6 |  | **−0.312** |  | 1.575 |
|  |  |  | 0.07* |  | 0.34* |
|  | 7 |  |  | **−0.358** | 0.974 |
|  |  |  |  | 0.07* | 0.18* |
| Logistic | 1 | −0.732 | 0.315 | 0.665 | −0.15 |
|  |  | 0.38 | 0.4 | 0.48 | 2.49 |
|  | 2 | **−0.864** | 0.537 |  | 0.802 |
|  |  | 0.36* | 0.35 |  | 0.8 |
|  | 3 | **−0.785** |  | 0.815 | 1.076 |
|  |  | 0.38* |  | 0.44 | 1.92 |
|  | 4 | **−1.07** |  |  | 3.947 |
|  |  | 0.35* |  |  | 3.95* |
|  | 5 |  | 0.5 | **0.93** | −4.606 |
|  |  |  | 0.39 | 0.41* | 1.65* |
|  | 6 |  | **0.963** |  | −4.901 |
|  |  |  | 0.32* |  | 1.48* |
|  | 7 |  |  | **1.171** | −2.98 |
|  |  |  |  | 0.37* | 0.88* |

## 4.4 Student Linearly Separable Data

### 4.4.1 Comparison of MNM and Nine "Diff1s"

We generate LSD from the original Student data by adding four to the study hour/day of the passing group. Table 4.7 lists a comparison of MNM and nine differences ("Diff1"). The four models include $X1$ and are linearly separable models. Six MP-based LDFs can discriminate these models correctly. On the other hand, most of the models of the four statistical discriminant functions cannot discriminate the four models correctly. Only two NMs of logistic regression are zero. Three models from $SN = 5$ to $SN = 7$ are not linearly separable. Two models of Revised LP-OLDF are not free from Problem 1. Because 11 "Diff1s" of the four statistical discriminant functions are negative values, the results might imply that they are not free from Problem 1.

### 4.4.2 Best Model

We examine the seven models listed in Table 4.8 by the Method 1. Four models are linearly separable models. Six MP-based LDFs and logistic regression can recognize that these four models are linearly separable. However, third and fourth "M1s" of logistic regression are not correct because two corresponding "Diff1s" in Table 4.7 are two and six. Four M1s and $M2$s of Revised IP-OLDF are zero. On the other hand, all $M1$s and $M2$s of Fisher's LDF are not zero. We compare Revised IP-OLDF with seven LDFs by the best model ($X1$). "$M2$Diffs" of SVM4, SVM1, LP, IPLP, H-SVM, logistic, and LDF are 0.55, 0.55, 0, 0, 0.63, 0.35, and 2.75 %, respectively. Although the amount of Student data is small, the Method 1 provides a wealth of information for researchers concerned with the study of a small sample.

Table 4.7 Comparison of MNM and nine "Diff1"

| SN | Var. | RIP | HSVM | SVM4 | SVM1 | LP | $f = 0$ | IPLP | Logistic | LDF | QDF | RDA |
|---|---|---|---|---|---|---|---|---|---|---|---|---|
| 1 | 1–3 | 0 | 0 | 0 | 0 | 0 | 0 | 0 | 0 | 2 | 2 | 0 |
| 2 | 1, 2 | 0 | 0 | 0 | 0 | 0 | 0 | 0 | 0 | 3 | 4 | 2 |
| 3 | 1, 3 | 0 | 0 | 0 | 0 | 0 | 0 | 0 | 2 | 3 | 3 | 3 |
| 4 | 1 | 0 | 0 | 0 | 0 | 0 | 0 | 0 | 6 | 6 | 6 | 6 |
| 5 | 2, 3 | 3 |  | 1 | 2 | 1 | 3 | 0 | −2 | 0 | −1 | −1 |
| 6 | 2 | 7 |  | 0 | 0 | −4 | 8 | 0 | −7 | −7 | −7 | −7 |
| 7 | 3 | 8 |  | 0 | 0 | 0 | 0 | 0 | −7 | −8 | −8 | −8 |

**Table 4.8** Best model

| RIP | | M1 | M2 | Var. | |
|---|---|---|---|---|---|
| 38s | 1 | 0 | 0 | 1, 2, 3 | |
| | 2 | 0 | 0 | 1, 2 | |
| | 3 | 0 | 0 | 1, 3 | |
| | 4 | 0 | **0** | 1 | |
| | 5 | 7.4 | 9.33 | 2, 3 | |
| | 6 | 14.98 | 17.5 | 2 | |
| | 7 | 17.9 | 21.33 | 3 | |
| SVM4 | | M1 | M2 | M1Diff. | M2Diff. |
| 38S | 1 | 0 | 1.75 | 0.00 | 1.75 |
| | 2 | 0 | 1.05 | 0.00 | 1.05 |
| | 3 | 0 | 1.30 | 0.00 | 1.30 |
| | 4 | 0 | **0.55** | 0.00 | **0.55** |
| | 5 | 10.63 | 12.8 | 3.23 | 3.48 |
| | 6 | 15.88 | 17.9 | 0.90 | 0.40 |
| | 7 | 19.55 | 20.4 | 1.65 | −0.93 |
| SVM1 | | M1 | M2 | M1Diff. | M2Diff. |
| 38S | 1 | 0 | 1.75 | 0.00 | 1.75 |
| | 2 | 0 | 1.05 | 0.00 | 1.05 |
| | 3 | 0 | 1.30 | 0.00 | 1.30 |
| | 4 | 0 | **0.55** | 0.00 | **0.55** |
| | 5 | 10.63 | 12.8 | 3.23 | 3.48 |
| | 6 | 15.88 | 17.9 | 0.90 | 0.40 |
| | 7 | 19.55 | 20.4 | 1.65 | −0.93 |
| LP | | M1 | M2 | M1Diff. | M2Diff. |
| 34S | 1 | 0 | 1.73 | 0.00 | 1.73 |
| | 2 | 0 | 1.08 | 0.00 | 1.08 |
| | 3 | 0 | 1.20 | 0.00 | 1.20 |
| | 4 | 0 | **0** | 0.00 | **0.00** |
| | 5 | 9.95 | 11.80 | 2.55 | 2.48 |
| | 6 | 11.43 | 13.00 | −3.55 | −4.50 |
| | 7 | 16.63 | 17.48 | −1.28 | −3.85 |
| IPLP | | M1 | M2 | M1Diff. | M2Diff. |
| 1m25S | 1 | 0 | 2.38 | 0.00 | 2.38 |
| | 2 | 0 | 0.83 | 0.00 | 0.83 |
| | 3 | 0 | 1.25 | 0.00 | 1.25 |
| | 4 | 0 | **0** | 0.00 | **0.00** |
| | 5 | 7.4 | 8.925 | 0.00 | −0.40 |
| | 6 | 14.98 | 17.5 | 0.00 | 0.00 |
| | 7 | 17.98 | 21.55 | 0.08 | 0.23 |

(continued)

## 4.4 Student Linearly Separable Data

| H-SVM | | M1 | M2 | M1Diff. | M2Diff. |
|---|---|---|---|---|---|
| 25s | 1 | 0 | 1.75 | 0.00 | 1.75 |
| | 2 | 0 | 1.05 | 0.00 | 1.05 |
| | 3 | 0 | 1.28 | 0.00 | 1.28 |
| | 4 | 0 | **0.63** | 0.00 | **0.63** |
| Logistic | | M1 | M2 | M1Diff. | M1Diff. |
| 4 m | 1 | 0 | 2.6 | 0.00 | 2.60 |
| | 2 | 0 | 1.225 | 0.00 | 1.23 |
| | 3 | 0 | 1.25 | 0.00 | 1.25 |
| | 4 | 0 | **0.35** | 0.00 | **0.35** |
| | 5 | 19.58 | 14.60 | 12.18 | 5.28 |
| | 6 | 32.70 | 17.27 | 17.73 | −0.23 |
| | 7 | 39.05 | 19.88 | 21.15 | −1.44 |
| LDF | | M1 | M2 | M1Diff. | M1Diff. |
| 1m30s | 1 | 7.70 | 4.02 | 7.70 | 4.02 |
| | 2 | 4.83 | 3.13 | 4.88 | 3.13 |
| | 3 | 5.03 | 3.12 | 5.03 | 3.12 |
| | 4 | 5.08 | **2.75** | 5.08 | **2.75** |
| | 5 | 23.8 | 16.04 | 0.28 | −4.15 |
| | 6 | 35.53 | 18.27 | −3.52 | −11.61 |
| | 7 | 39.13 | 19.86 | −5.28 | −14.92 |

### 4.4.3 95 % CI of Discriminant Coefficient

Table 4.9 lists the 95 % CI of the discriminant coefficient. We compare four linearly separable models: (X1, X2, X3), (X1, X2), (X1, X3), and (X1). Although the best model is a one-variable model (X1), we examine four models. The four Revised IP-OLDFs become similar to LDF in Eq. (4.1). These results show that all 100 LDFs are the same, and the two coefficients of X2 and X3 are zero. Therefore, all 400 Revised IP-OLDFs select the same linear discriminant hyperplane X1 = 5.5. Revised IP-OLDF indicates that discrimination over two variables does not need to be considered. Moreover, it can select variables naturally.

$$\text{RIP} = 2 \times X1 - 11 \tag{4.1}$$

Because other coefficients of the five MP-based LDFs vary, the three numeric values with separated slash (/) are 2.5, 50 (median), and 97.5 %. All the coefficients of X2 and X3 of the other five LDFs are zero in the three- and two-variable models because 95 % CI includes zero. All coefficients of X1 are positive.

**Table 4.9** 95 % CI of discriminant coefficient

|        | X1          | X2            | X3            | C            |
|--------|-------------|---------------|---------------|--------------|
| RIP    | 2           | 0             | 0             | −11          |
| HSVM   | 0.5/**1.1**/1.9 | −0.7/0.2/0.4 | −0.7/−0.4/0.3 | −10.4/−6/1   |
| SVM4   | 0.5/**1.1**/1.9 | −0.7/0.2/0.4 | −0.7/−0.4/0.3 | −10.4/−6/1   |
| SVM1   | 0.5/**1**/1.9 | −0.7/0.1/0.4  | −0.7/−0.2/0.3 | −10.4/−6/1   |
| IPLP   | 0.1/**1.1**/2.2 | −1.7/0.2/1.1 | −1.7/−0.6/1.7 | −11.6/−6.3/5.2 |
| LP     | 0.5/**1.1**/2 | −0.9/0.2/1.1  | −1.7/−0.5/0.5 | −11/−6/0.2   |
| RIP    | 2           | 0             |               | −11          |
| HSVM   | 0.6/**1.2**/2 | −1/0/0.4      |               | −11/−**6**/−1 |
| SVM4   | 0.6/**1.2**/2 | −0.1/0/0.4    |               | −11/−**6**/−1 |
| SVM1   | 0.6/**1**/2 | −0.1/0/0.4    |               | −11/−**6**/−1 |
| IPLP   | 0.6/**1.33**/3.3 | −1.7/0/1.3  |               | −24.3/−**8.3**/−0.5 |
| LP     | 0.6/**1.2**/2 | −1/0/0.4      |               | −11/−**6**/−0.5 |
| RIP    | 2           |               | 0             | −11          |
| HSVM   | 0.6/**1**/2 |               | −1/0/0.7      | −11/−**6**/−2.8 |
| SVM4   | 0.6/**1**/2 |               | −0.1/0/0.7    | −11/−**6**/−2.8 |
| SVM1   | 0.6/**1**/2 |               | −1/0/0.7      | −11/−**6**/−2.8 |
| IPLP   | 0.6/**1**/6 |               | −1/0/4        | −43/−**7.5**/−2.4 |
| LP     | 0.6/**1**/2 |               | −1/0/0.7      | −11/−**6.5**/−2.2 |
| RIP    | 2           |               |               | −11          |
| HSVM   | 0.7/**2**/2 |               |               | −11/−**11**/−3.7 |
| SVM4   | 0.7/**2**/2 |               |               | −11/−**11**/−3.7 |
| SVM1   | 0.7/**1**/2 |               |               | −11/−**6**/−3.7 |
| IPLP   | 0.7/**2**/2 |               |               | −11/−**11**/−3.7 |
| LP     | 0.7/**2**/2 |               |               | −11/−**11**/−3.7 |

The regression coefficients of Fisher's LDF are in Eq. (4.2). The logistic regression is in the Eq. (4.3). The numbers in parentheses are SEs. All coefficients are rejected at 5 % level. SEs of logistic regression are enormous (Firth 1993), and all CIs include zero.

$$\text{LDF} = 0.18 \times X1 - 0.05 \times X2 - 0.015 \times X3 - 0.9 \atop (0.003) \qquad (0.007) \qquad (0.007)\,(0.04)$$

(4.2)

$$\text{Logi} = -31 \times X1 - 11 \times X2 + 17 \times X3 + 182 \atop (1123) \qquad (1420) \qquad (2068)\,(7227)$$

(4.3)

## 4.5 Summary

Many statisticians believe that the MNM criterion is an irrational criterion because it over fits training samples and overestimate validation samples. On the contrary, the generalization ability of LDF is best because it follows the normal distribution without examination by real data. In this book, we prove that the claim is wrong through many results of the best model for Revised IP-OLDF and Fisher's LDF. In addition, the mean error rates, $M1$ and $M2$, from the training and validation samples of Fisher's LDF, respectively, are higher than other LDFs. Previous critical studies that have used LDF should be reviewed, especially in medical diagnosis. Method 1 is very useful compared with LOO method. Moreover, Student linearly separable data explain why Revised IP-OLDF can select features naturally.

## References

Anderson E (1945) The irises of the Gaspe Peninsula. Bull Am Iris Soc 59:2–5
Cox DR (1958) The regression analysis of binary sequences (with discussion). J Roy Stat Soc B 20:215–242
Firth D (1993) Bias reduction of maximum likelihood estimates. Biometrika 80:27–39
Fisher RA (1936) The use of multiple measurements in taxonomic problems. Ann Eugenics 7:179–188
Fisher RA (1956) Statistical methods and statistical inference. Hafner Publishing Co, New Zealand
Flury B, Rieduyl H (1988) Multivariate statistics: a practical approach. Cambridge University Press, New York
Friedman JH (1989) Regularized discriminant analysis. J Am Stat Assoc 84(405):165–175
Ishii A, Yata K, Aoshima M (2014) Asymptotic distribution of the largest eigenvalue via geometric representations of high-dimension, low-sample-size data. Sri Lankan J Appl Statist, special issue: modern statistical methodologies in the cutting edge of science (ed. Mukhopadhyay, N.): 81–94
Jeffery IB, Higgins DG, Culhane C (2006) Comparison and evaluation of methods for generating differentially expressed gene lists from microarray data. BMC Bioinf. Jul 26 7:359:1–16. doi: 10.1186/1471-2105-7-359
Miyake A, Shinmura S (1976) Error rate of linear discriminant function. In: Dombal FT, Gremy F (ed). North-Holland Publishing Company, The Netherland, pp 435–445
Miyake A, Shinmura S (1979) An algorithm for the optimal linear discriminant functions. Proceedings of the international conference on cybernetics and society, pp 1447–1450
Miyake A, Shinmura S (1980) An algorithm for the optimal linear discriminant function and its application. Jpn Soc Med Electron Biol Eng 18(6):452–454
Sall JP (1981) SAS regression applications. SAS Institute Inc., USA. (Shinmura S. translate Japanese version)
Sall JP, Creighton L, Lehman A (2004) JMP start statistics, third edition. SAS Institute Inc.. USA. (Shinmura S. edits Japanese version)
Schrage L (1991) LINDO—an optimization modeling systems. The Scientific Press, UK. (Shinmura S. & Takamori, H. translate Japanese version)
Schrage L (2006) Optimization modeling with LINGO. LINDO Systems Inc., USA. (Shinmura S. translates Japanese version)
Shinmura S, Miyake A (1979) Optimal linear discriminant functions and their application. COMPSAC 79:167–172

Shinmura S (1998) Optimal linear discriminant functions using mathematical programming. J Jpn Soc Comput Stat, 11/2:89–101

Shinmura S, Tarumi T (2000) Evaluation of the optimal linear discriminant functions using integer programming (IP-OLDF) for the normal random data. J Jpn Soc Comput Stat 12(2):107–123

Shinmura S (2000a) A new algorithm of the linear discriminant function using integer programming. New Trends Prob Stat 5:133–142

Shinmura S (2000b) Optimal linear discriminant function using mathematical programming. Dissertation, 200:1–101, Okayama University, Japan

Shinmura S (2003) Enhanced algorithm of IP-OLDF. ISI2003 CD-ROM, pp 428–429

Shinmura S (2004) New algorithm of discriminant analysis using integer programming. IPSI 2004 Pescara VIP Conference CD-ROM, pp 1–18

Shinmura S (2005) New age of discriminant analysis by IP-OLDF—Beyond Fisher's linear discriminant function. ISI2005, pp 1–2

Shinmura S (2007) Overviews of discriminant function by mathematical programming. J Jpn Soc Comput Stat 20(1–2):59–94

Shinmura S (2010a) The optimal linearly discriminant function (Saiteki Senkei Hanbetu Kansuu). Union of Japanese Scientist and Engineer Publishing, Japan

Shinmura S (2010c) Improvement of CPU time of revised IP-OLDF using linear programming. J Jpn Soc Comput Stat 22(1):39–57

Shinmura S (2011) Problems of discriminant analysis by mark sense test data. Jpn Soc Appl Stat 40(3):157–172

Shinmura S (2013) Evaluation of optimal linear discriminant function by 100-fold cross-validation ISI CD-ROM, pp 1–6

Shinmura S (2014a) End of discriminant functions based on variance-covariance matrices. ICORE2014, pp 5–16

Shinmura S (2014b) Improvement of CPU time of linear discriminant functions based on MNM criterion by IP. Stat Optim Inf Comput 2:114–129

Shinmura S (2014c) Comparison of linear discriminant functions by $K$-fold cross-validation. Data Anal 2014:1–6

Shinmura S (2015a) The 95 % confidence intervals of error rates and discriminant coefficients. Stat Optim Inf Comput 2:66–78

Shinmura S (2015b) A trivial linear discriminant function. Stat Optim Inf Comput 3:322–335. doi:10.19139/soic.20151202

Shinmura S (2015c) Four serious problems and new facts of the discriminant analysis. In: Pinson E, Valente F, Vitoriano B (ed) Operations research and enterprise systems, pp 15–30. Springer, Berlin (ISSN: 1865-0929, ISBN: 978-3-319-17508-9. doi:10.1007/978-3-319-17509-6)

Shinmura S (2015d) Four problems of the discriminant analysis. ISI 2015:1–6

Shinmura S (2016a) The best model of Swiss banknote data. Stat Optim Inf Comput, 4: 118–131. International Academic Press (ISSN: 2310-5070 (online) ISSN: 2311-004X (print), doi:10.19139/soic.v4i2.178)

Shinmura S (2016b) Matroska feature selection method for microarray data. Biotechnology 2016:1–8

Shinmura S (2016c) Discriminant analysis of the linear separable data—Japanese-automobiles. J Stat Sci Appl X, X: 0–14

Simon N, Friedman J, Hastie T, Tibshirani R (2013) A sparse-group lasso. J Comput Graph Stat 22:231–245

Vapnik V (1995) The nature of statistical learning theory. Springer, Berlin

# Chapter 5
# Pass/Fail Determination Using Examination Scores

## A Trivial Linear Discriminant Function

## 5.1 Introduction

In this chapter, we examine the $k$-fold cross-validation for small sample method (Method 1) by combining the resampling technique with $k$-fold cross-validation (Shinmura 2010a, 2013, 2014c, 2015a, b, c, 2016a, b, c). By this breakthrough, we obtain the error rate means, $M1$ and $M2$, for the training and validation samples, respectively, and the 95 % CI of the discriminant coefficient and the error rate (Miyake and Shinmura 1976). Fisher (1936, 1956) described Fisher's linear discriminant function (LDF) and founded the discriminant theory. He never formulated SE of Fisher's LDF (Problem 4) (Shinmura, 2014a, 2015c-d). Therefore, there was no sophisticated model selection procedure for the discriminant analysis. Although Lachenbruch and Mickey (1968) proposed LOO procedure for model selection of the discriminant analysis, they could not achieve the new procedure because of lack of computer power. If we set $k = 100$, we can obtain 100 LDFs and 100 error rates for the training and validation samples. From the 100 error rates, we calculate two means, $M1$ and $M2$, for the training and validation samples, respectively. We consider the model with minimum M2 among all possible combination models (Goodnight 1978) as the best model. We apply Method 1 and procedure for six datasets of the pass/fail determinations using examination scores and obtain good results. We should distinguish these computer-intensive approaches from the traditional inferential statistics with SE. Genuine statisticians without computer power established inferential statistics intellectually. Currently, we can utilize computer power, statistical software such as JMP (Sall et al. 2004), and an MP solver such as LINGO (Schrage 1991, 2006). Therefore, we propose the Theory by computer-intensive approach. Those researchers who want to discriminate their research data can determine the best model in addition to the 95 % CI of the error

© Springer Science+Business Media Singapore 2016
S. Shinmura, *New Theory of Discriminant Analysis After R. Fisher*,
DOI 10.1007/978-981-10-2164-0_5

rate and discriminant coefficients (Shinmura 2010a). Method 1 and the best model provide precise and deterministic judgment with regard to the evaluation and validation of six MP-based LDFs and two statistical LDFs. Six MP-based LDFs are Revised IP-OLDF (Shinmura 1998, 2003, 2004, 2005, 2007, 2011a, b), Revised LP-OLDF, Revised IPLP-OLDF (Shinmura 2010b, 2014b), hard-margin support vector machine (H-SVM) (Vapniik 1995), and two soft-margin SVMs, such as SVM4 and SVM1. Two statistical LDFs are Fisher's LDF and logistic regression (Cox 1958). Moreover, we obtain the actual results of the discriminant coefficients by setting the intercept of MP-based LDFs to one. Although we could obtain 95 % CI of the discriminant coefficients in 2000, we could not explain the useful meaning of 95 % CI of the discriminant coefficients clearly (Shinmura 2015a). By setting the intercept to one, most LDFs, with the exception of Fisher's LDF, are almost the same as trivial LDFs (Shinmura 2015b).

## 5.2 Pass/Fail Determination Using Examination Scores Data in 2012

After 2010, we taught a preliminary statistical course to approximately 130 first-year students. Midterm and final examinations consisted of 100 questions with ten choices. We consider discrimination using four testlet scores as independent variables (Shinmura 2011b). If the passing mark is 50 points, we can easily obtain a trivial LDF ($f = T1 + T2 + T3 + T4 - 50$) with NM of zero. If $f \geq 0$ or $f < 0$, the students pass or fail the examination, respectively. Usually, all LDFs, with the exception of Revised IP-OLDF, are not free from Problem 1. Because we can define the discriminant rule by examination scores (or independent variables), we can obtain the aforementioned trivial LDF that is free from Problem 1. We propose that the pass/fail determination using examination scores provides deep knowledge with regard to discrimination and offers useful test data for linearly separable data (LSD). We wrote many papers on medical diagnosis (Nomura and Shinmura 1978; Shimize et al. 1975; Shinmura et al. 1983, 1987; Shinmura 1984). However, we have no connection with medical doctors now. Therefore, we use these data instead of medical data because both data have many cases nearby the discriminant hyperplane. Table 5.1 lists the discrimination of four testlet scores for 10, 50, and 90 % levels of the midterm examinations from 2012. A total of 124 students attended the examination. "$p$" denotes the number of independent variables selected by the forward stepwise technique. At the 10 % level, the two-variable model ($T4$, $T2$) is linearly separable. There are 15 discriminant models by all the combinations of four variables, only four LDFs of which are linearly separable. The passing mark is 36 points, and ten students failed the examination. At the 50 % level, the only full model is linearly separable. The passing mark is 63 points, and 57 students failed the examination. At the 90 % level, the only full model is linearly separable. The

## 5.2 Pass/Fail Determination Using Examination Scores Data in 2012

**Table 5.1** NMs of four discriminant functions by forward stepwise in 2012 midterm examinations

|       | p | Var. | RIP | Logistic | LDF[a] | QDF[a] |
|-------|---|------|-----|----------|--------|--------|
| 10 %  | 1 | 4    | 4   | 8        | 6      | 6      |
|       | 2 | 2    | **0** | **0**  | **1**  | **1**  |
|       | 3 | 1    | 0   | 0        | 1      | 1      |
|       | 4 | 3    | 0   | 0        | 1      | 1      |
| 50 %  | 1 | 4    | 12  | 12       | 14     | 14     |
|       | 2 | 1    | 6   | 5        | 9      | 9      |
|       | 3 | 2    | 3   | 3        | 8      | 8      |
|       | 4 | 3    | **0** | **0**  | 7      | 7      |
| 90 %  | 1 | 3    | 8   | 30       | 12     | 12     |
|       | 2 | 1    | 5   | 12       | 9      | 9      |
|       | 3 | 4    | 3   | 3        | 10     | 10     |
|       | 4 | 2    | **0** | **0**  | 10     | 10     |

[a]We obtained some NMs of Fisher's LDF and QDF by two options of "prior probability = Same" or "prior probability = Proportional" using JMP to this point. After this paper (Shinmura 2015b), we set the option to "prior probability = Proportional."

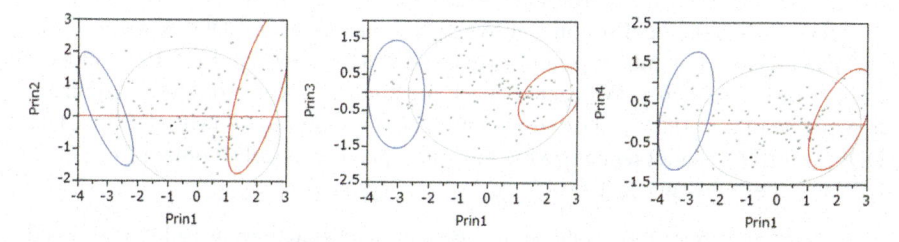

**Fig. 5.1** Scatter plots for three groups

passing mark is 78 points, and 112 students failed the examination. "RIP and Logistic" are Revised IP-OLDF and logistic regression. Both LDFs can discriminate a linearly separable model correctly. However, logistic regression cannot sometimes discriminate LSD correctly. On the other hand, Fisher's LDF and QDF cannot discriminate all linearly separable models in these data.

Figure 5.1 shows three scatter plots by PCA. The x-axis is the first principal component. The y-axes correspond to the second, third, and fourth principal components, from left to right. Three 95 % probability ellipses correspond to three groups, such as SCORE $\leq$ 35, 36 $\leq$ SCORE $\leq$ 77, and 78 $\leq$ SCORE. The three groups consist of 10, 102, and 12 students. The first and third groups are almost symmetrical because the two ellipses and cases are almost the same. If we check both NMs of the full model at the three levels of Table 5.1, such NMs for the Fisher's LDF and QDF are 1, 7, and 10, respectively, for each level. We cannot explain whether the increasing trend is common for other data.

## 5.3 Pass/Fail Determination by Examination Scores (50 % Level in 2012)

In this section, we discuss discrimination at the 50 % level. The passing mark is 63 points. Table 5.1 indicates that only the full model is a linearly separable model. We know that trivial LDF is $f = T1 + T2 + T3 + T4 - 63$.

### 5.3.1 MNM and Nine NMs

Table 5.2 lists the MNM (Miyake and Shinmura 1979, 1980; Shinmura and Miyake 1979; Shinmura and Tarumi 2000; Shinmura 2000a, b) and nine "Diff1" of seven LDFs and two discriminant functions. We omit four one-variable model because there is no meaning for discrimination. The seven LDFs are as follows: H-SVM, SVM4, SVM1, Revised LP-OLDF (LP), Revised IPLP-OLDF (IPLP), logistic regression (Logistic), and Fisher's LDF (LDF). The two discriminant functions are QDF and RDA (Friedman 1989). Nine "Diff1" are the difference defined as (nine NMs − MNM). The first column is the sequential number (SN) of 11 models from four to two variables that correspond to the second column, "Model." We check the number of cases on the discriminant hyperplane of the seven LDFs. We show it in the parenthesized numbers. Only five two-variable Revised LP-OLDF models cannot avoid Problem 1. We cannot check the number of cases on the discriminant hyperplane of four statistical discriminant functions analyzed by JMP. SVM1, LDF, QDF, and RDA cannot recognize the linearly separable model (Problem 2). "Diff1" indicates the following facts:

1. We can approximately evaluate nine discriminant functions as follows: Revised IPLP-OLDF is the best result because ten NMs of IPLP are the same as MNM. Logistic regression is the second best because four NMs of logistic regression

**Table 5.2** MNM and nine "Diff1" (50 % level)

| SN | Model | RIP | H-SVM | SVM4 | SVM1 | LP | IPLP | Logistic | LDF | QDF | RDA |
|---|---|---|---|---|---|---|---|---|---|---|---|
| 1 | 1–4 | 0 | 0 | 0 | 3 | 0 | 0 | 0 | 7 | 5 | 4 |
| 2 | 1, 2, 4 | 3 | | 1 | 1 | 1 | 0 | 0 | 5 | 5 | 3 |
| 3 | 2–4 | 5 | | 3 | 2 | 3 | 0 | 0 | 5 | 5 | 3 |
| 4 | 1, 3, 4 | 3 | | 0 | 0 | 0 | 0 | 1 | 6 | 2 | 2 |
| 5 | 1–3 | 13 | | 4 | 5 | 4 | 0 | 1 | 4 | 3 | 2 |
| 6 | 2, 4 | 6 | | 1 | 1 | −2(6) | 0 | 1 | 7 | 4 | 6 |
| 7 | 1, 4 | 5 | | 2 | 2 | 2 | 0 | 0 | 4 | 3 | 3 |
| 8 | 3, 4 | 10 | | 1 | 2 | −2(6) | 0 | 1 | 1 | 1 | 0 |
| 9 | 1, 2 | 16 | | 2 | 3 | 0(6) | 0 | 1 | 4 | 5 | 5 |
| 10 | 2, 3 | 26 | | 6 | 6 | 2(11) | 3 | 1 | 7 | 7 | 7 |
| 11 | 1, 3 | 17 | | 2 | 1 | 0(3) | 0 | 1 | 2 | 1 | 1 |

5.3 Pass/Fail Determination by Examination Scores (50 % Level in 2012) 103

are the same as MNM. Three statistical discriminant functions, with the exception of logistic regression, are the worst results.

2. Because five LP models have several cases on the discriminant hyperplane and two "Diff1s," are negative, we cannot find Problem 1 by checking negative "Diff1" models. We expect statistical software to output this number for users.

### 5.3.2 Error Rate Means (M1 and M2)

Table 5.3 lists the results by Method 1. We omit QDF and RDA because those are not LDFs. We examine 11 discriminant models for seven LDFs and only the full model for H-SVM. The first 11 rows are 11 Revised IP-OLDF (RIP) models. $M1$s and $M2$s are the error rate means for the training and validation samples, respectively. We confirm that the full models for the eight LDFs are the best models because the $M2$ values are the minimum among 11 models (Shinmura 2016a, 2016c). The eight $M2$s of the best models are 1, 1.09, 1.09, 1.79, 1.02, 1.06, 0.97, and 5.58 %. "$M1$Diff and $M2$Diff" are the differences defined as ($M1$ and $M2$ of seven LDFs – those of RIP). The minimum "$M2$Diff" of the seven LDFs are 0.09, 0.09, 0.79, 0.02, 0.06, −0.03, and 4.58%. $M2$ of Fisher's LDF is 4.58 % greater than that of Revised IP-OLDF. Although we cannot evaluate the influence of Problem 1, the best logistic model is 0.03 % less than that of Revised IP-OLDF. However, this good result of logistic regression may be caused by Problem 1 and ignores the cases on the discriminant hyperplane. The other five LDFs are acceptable because "$M2$Diff" is within 0.79 % higher than that of Revised IP-OLDF. The column "Diff" is the difference between error rate means defined as ($M2 - M1$). Some statisticians claim that the model with a minimum value of "Diff" has good generalization ability. If this claim is correct, the sixth model ($T1$, $T2$, $T3$) of Fisher's LDF has good generalization ability. $M1$ and $M2$ of this model are very high: 9.03 and 9.15 %, respectively. Therefore, we cannot permit this claim. If we observe the second best model among those ten models that are not linearly separable, only Fisher's LDF selects the fourth model, whose $M2$ is 5.60%. On the other hand, the other seven LDFs accept the second model. The second "$M2$Diff" of six LDFs are 0.41, 0.56, 0, 0.35, 0.27, and 3.03 %. This fact might imply that Revised IP-OLDF is superior to other LDFs for nonlinearly separable models, but not for linearly separable models.

### 5.3.3 95 % CI of Discriminant Coefficients

We examine the 95 % CI of the best models. Table 5.4 lists the median and 95 % CI of six MP-based LDFs by setting the intercept of MP-based LDFs to one. It is our mistake that Fisher's LDF and the logistic regression by the JMP script

**Table 5.3** M1s and M2s of eight LDFs (50 %)

| RIP | | M1 | M2 | Diff. | Model | |
|---|---|---|---|---|---|---|
| | 1 | 0 | **1.00** | **1.00** | 1, 2, 3, 4 | |
| | 2 | 2 | **3.12** | 1.45 | 1, 2, 4 | |
| | 3 | 3 | 5.66 | 2.46 | 1, 3, 4 | |
| | 4 | 1 | 3.36 | 1.86 | 2, 4 | |
| | 5 | 9.13 | 12.57 | 3.44 | 1, 3, 4 | |
| | 6 | 4.51 | 5.57 | 1.07 | 1, 2, 3 | |
| | 7 | 2.98 | 4.40 | 1.41 | 1, 4 | |
| | 8 | 6.24 | 9.16 | 2.92 | 3, 4 | |
| | 9 | 11.75 | 14.35 | 2.60 | 1, 2 | |
| | 10 | 19.42 | 22.16 | 2.74 | 2, 3 | |
| | 11 | 12.18 | 14.58 | 2.40 | 1, 3 | |
| H-SVM | | *M*1 | *M*2 | Diff. | *M*1Diff. | *M*2Diff. |
| | 1 | 0.00 | **1.09** | 1.09 | 0.00 | **0.09** |
| SVM4 | 1 | 0.00 | **1.09** | 1.09 | 0.00 | **0.09** |
| | 2 | 2.61 | **3.53** | 0.92 | 0.94 | **0.41** |
| | 9 | 14.57 | 15.21 | **0.65** | 2.82 | 0.86 |
| SVM1 | 1 | 0.47 | **1.79** | 1.32 | 0.47 | **0.79** |
| | 2 | 2.82 | **3.68** | 0.86 | 1.15 | **0.56** |
| | 7 | 4.38 | 5.06 | **0.68** | 1.40 | 0.66 |
| IPLP | 1 | 0.00 | **1.02** | **1.02** | 0.00 | **0.02** |
| | 2 | 1.67 | **3.12** | 1.45 | 0.00 | **0.00** |
| LP | 1 | 0.00 | **1.06** | 1.06 | 0.00 | **0.06** |
| | 2 | 2.59 | **3.47** | 0.88 | 0.92 | **0.35** |
| | 9 | 13.91 | 14.56 | **0.65** | 2.16 | 0.21 |
| Logistic | 1 | 0.00 | **0.97** | 0.97 | 0.00 | **−0.03** |
| | 2 | 2.60 | **3.39** | 0.79 | 0.93 | **0.27** |
| | 9 | 14.31 | 14.60 | **0.30** | 2.56 | 0.26 |
| LDF | 1 | 5.15 | **5.58** | 0.43 | 5.15 | **4.58** |
| | 2 | 5.65 | **6.15** | 0.49 | 3.98 | **3.03** |
| | 4 | 5.40 | **5.60** | 0.21 | 3.90 | **2.25** |
| | 6 | 9.03 | **9.15** | **0.12** | 4.52 | 3.58 |

designed by us do not output 100 discriminant coefficients. If the 95 % CI includes zero, we can determine that the pseudo-population coefficient is zero. If the value of 2.5 % is greater than zero or the value of 97.5 % is less than zero, we estimate the pseudo-population coefficient as a positive or negative value. Following this determination, only five $T2$ coefficients, with the exception of SVM1, are zero, and the other coefficients are negative. The four medians of the full model are almost −0.016. This fact implies that these LDFs are the same as the trivial LDF in Eq. (5.1):

## 5.3 Pass/Fail Determination by Examination Scores (50 % Level in 2012)

**Table 5.4** 95 % CI of six LDFs

|  |  | T1 | T2 | T3 | T4 | c |
|---|---|---|---|---|---|---|
| RIP | 97.5 % | −0.0075 | 0.0082 | −0.0069 | −0.0117 | 1 |
|  | Median | **−0.0160** | **−0.0160** | −0.0161 | −0.0162 | 1 |
|  | 2.5 % | −0.0204 | −0.0243 | −0.0262 | −0.0246 | 1 |
| H-SVM | 97.5 % | −0.0075 | 0 | −0.0081 | −0.0120 | 1 |
|  | Median | **−0.0160** | **−0.0160** | **−0.0160** | **−0.0160** | 1 |
|  | 2.5 % | −0.0196 | −0.0217 | −0.0258 | −0.0232 | 1 |
| SVM4 | 97.5 % | −0.0075 | 0 | −0.0081 | −0.012 | 1 |
|  | Median | **−0.0160** | **−0.0160** | **−0.0160** | **−0.0160** | 1 |
|  | 2.5 % | −0.0196 | −0.0217 | −0.0258 | −0.0232 | 1 |
| SVM1 | 97.5 % | −0.0106 | −0.0010 | −0.0085 | −0.0121 | 1 |
|  | Median | −0.0154 | −0.0161 | −0.0164 | −0.0171 | 1 |
|  | 2.5 % | −0.0202 | −0.0230 | −0.0250 | −0.0208 | 1 |
| IPLP | 97.5 % | −0.0080 | 0.0071 | −0.0069 | −0.0118 | 1 |
|  | Median | **−0.0160** | **−0.0160** | **−0.0160** | **−0.0160** | 1 |
|  | 2.5 % | −0.0204 | −0.0239 | −0.0272 | −0.0243 | 1 |
| LP | 97.5 % | −0.0095 | 0.0071 | −0.0082 | −0.0117 | 1 |
|  | Median | **−0.0160** | **−0.0160** | **−0.0160** | **−0.0160** | 1 |
|  | 2.5 % | −0.0204 | −0.0243 | −0.0264 | −0.0234 | 1 |

$$F = -0.016 \times T1 - 0.016 \times T2 - 0.016 \times T3 - 0.016 \times T4 + 1$$
$$= -T1 - T2 - T3 - T4 + 62.5 \quad (5.1)$$

We can find this surprising result by setting the intercept to one. When we did not set the intercept, we could not find useful meaning of discriminant coefficient (Shinmura 2015a). We discriminate the pseudo-population sample that has 12,400 cases by Fisher's LDF and logistic regression. Equation (5.2) is a logistic regression. If we divide five coefficients by 2178, we can obtain the same trivial LDF. Because the numbers in parentheses are SEs, all coefficients are zero.

$$\text{Logist 1234} = -34.33 \times T1 - 35.07 \times T2 - 35.27 \times T3 - 35.14 \times T4 + 2178$$
$$(5062) \quad (4631) \quad (4899) \quad (4607)(277,555)$$
$$= -0.016 \times T1 - 0.016 \times T2 - 0.016 \times T3 - 0.016 \times T4 + 1 \quad (5.2)$$

On the other hand, we obtain Fisher's LDF in Eq. (5.3). We obtain these coefficients by regression analysis (Sall 1981; Schrage 1991). If we divide the coefficients by −3.22, we can determine that Fisher's LDF is not the same as the trivial LDF because the third coefficient becomes −0.008. This fact indicates that Fisher's LDF does not follow the real data, but assumes that the data are the normal distribution. The numbers in parentheses are SEs, and we know that all coefficients are rejected at 5 %.

$$\begin{aligned}\text{LDF } 1234 &= 0.059 \times T1 + 0.056 \times T2 + 0.026 \times T3 + 0.056 \times T4 - 3.22\\&\quad(0.001^*) \quad\quad (0.003^*) \quad\quad (0.002^*) \quad\quad (0.0007^*)(0.026^*)\\&= -0.018 \times T1 - 0.017 \times T2 - 0.008 \times T2 - 0.017 \times T4 + 1\end{aligned}$$

(5.3)

## 5.4 Pass/Fail Determination by Examination Scores (90 % Level in 2012)

In this section, we discuss discrimination at the 90 % level. The passing mark is 78 points. Table 5.1 indicates that only the full model is a linearly separable model. We know that trivial linearly separable LDF is $f = T1 + T2 + T3 + T4 - 78$ (or 77.5).

### 5.4.1 MNM and Nine NMs

Table 5.5 lists MNM and nine "Diff1" of eight LDFs and two discriminant functions. We check the number of cases on the discriminant hyperplane of six MP-based LDFs. The three Revised LP-OLDF models cannot avoid Problem 1. We cannot check the number of cases on the discriminant hyperplane of four statistical discriminant functions because statistical software companies do not know Problem 1. The three statistical discriminant functions and SVM1 cannot recognize the linearly separable model (Problem 2). "Diff1" indicates the following facts:

1. We can roughly evaluate nine discriminant functions as follows. Revised IPLP-OLDF is the best because ten NMs of IPLP are the same as MNM.
2. SVM1, LDF, RDA, and QDF cannot discriminate the linearly separable model exactly (Problem 2).
3. Although logistic regression is the second best at the 50 % level, ten NMs of nonlinearly separable models are greater than MNM. Moreover, logistic regression is worse than SVM4, SVM1, LP, and IPLP. This result is crucial because logistic regression is as same as IPLP and is better than SVM4, SVM1, and LP in other data. Although logistic regression is most reliable among statistical discriminant functions, these results show the defect of logistic regression.
4. Three statistical discriminant functions are worse than RIP and IPLP. Although we shall never use these discriminant functions anymore, statistical software companies shall support these functions because these functions are our heritage and we need the education of discriminant analysis.

## 5.4 Pass/Fail Determination by Examination Scores (90 % Level in 2012)

**Table 5.5** MNM and nine "Diff1" (90 % level)

| SN | Model | RIP | H-SVM | SVM4 | SVM1 | LP | IPLP | Logistic | LDF | QDF | RDA |
|---|---|---|---|---|---|---|---|---|---|---|---|
| 1 | 1–4 | 0 | 0 | 0 | 1 | 0 | 0 | 0 | 10 | 1 | 10 |
| 2 | 1, 2, 4 | 3 |  | 1 | 3 | 1 | 1 | 3 | 9 | 6 | 9 |
| 3 | 2–4 | 5 |  | 1 | 0 | 0(1) | 0 | 2 | 4 | 3 | 4 |
| 4 | 1, 3, 4 | 2 |  | 1 | 1 | 1 | 0 | 1 | 8 | 5 | 8 |
| 5 | 1–3 | 5 |  | 2 | 2 | 2 | 0 | 2 | 5 | 2 | 5 |
| 6 | 2, 4 | 5 |  | 2 | 2 | 2 | 0 | 10 | 7 | 2 | 7 |
| 7 | 1, 4 | 5 |  | 3 | 3 | 3 | 0 | 5 | 7 | 1 | 7 |
| 8 | 3, 4 | 6 |  | 4 | 5 | 4(1) | 0 | 5 | 3 | 5 | 3 |
| 9 | 1, 2 | 11 |  | 1 | 1 | 1 | 0 | 27 | 1 | 6 | 1 |
| 10 | 2, 3 | 8 |  | 0 | 2 | 0(2) | 0 | 32 | 2 | 3 | 2 |
| 11 | 1, 3 | 7 |  | 2 | 2 | 2 | 0 | 5 | 2 | 2 | 2 |

5. We find the Problem 1 of Revised LP-OLDF for three models (SN = 3, 8, 10) with non-negative "Diff1s".

### 5.4.2 Error Rate Means (M1 and M2)

Table 5.6 lists the results by Method 1. SVM1 and Fisher's LDF cannot recognize that the full model is linearly separable. In particular, $M1$ of Fisher's LDF is 11.61 %. Only Fisher's LDF selects the fourth model as the best model. The other LDFs select the full model as the best model. The seven "$M2$Diff" of the full model are −0.03, −0.03, 0.16, −0.02, −0.05, 0.08, and 11.53 %. Although we cannot evaluate the influence of Problem 1, the four best models of H-SVM, SVM4, IPLP, and LP are better than those of Revised IP-OLDF within 0.05 %. This fact implies the following.

$M2$ of Revised IP-OLDF might be wrong for linearly separable models. On the other hand, the best model for Fisher's LDF is 11.525 % worse than RIP. If we look at the best model among those ten that are not linearly separable, all LDFs select a fourth model. The six "$M2$Diff" of the fourth model are 0.29, 0.78, 0.04, 0.23, 0.33, and 10.25 %. Five LDFs, with the exception of Fisher's LDF, are acceptable. Only Fisher's LDF is 10.25 % worse than Revised IP-OLDF. If we compare two best models among 11 and ten models, Revised IP-OLDF is superior to other LDFs for nonlinearly separable models, but not for linearly separable models. The sixth "Diff" of Fisher's LDF is the minimum value −0.15 %. This fact means that the sixth $M2$ (26.72 %) is 0.15 % less than $M1$ (26.87 %). Some statisticians claim that Fisher's LDF has good generalization ability in this model. Our research shows that this claim is entirely wrong.

**Table 5.6** *M*1s and *M*2s of eight LDFs (90 %)

|  |  | M1 | M2 | Diff. | Model |  |
|---|---|---|---|---|---|---|
| RIP | 1 14m8s | 0 | **1.06** | 1.06 | 1, 2, 3, 4 |  |
|  | 4 | 0.61 | **2.32** | 1.70 | 1, 3, 4 |  |
|  | 6 | 3.96 | 4.69 | **0.73** | 2, 4 |  |
|  |  | M1 | M2 | Diff. | M1Diff. | M2Diff. |
| H-SVM | 1 5m23s | 0 | **1.02** | 1.02 | 0 | **−0.03** |
| SVM4 | 1 9m23s | 0 | **1.02** | 1.02 | 0 | **−0.03** |
|  | 4 | 1.07 | **2.61** | 1.53 | 0.46 | **0.29** |
|  | 11 | 6.28 | 6.95 | **0.67** | 1.90 | 0.46 |
| SVM1 | 1 8m53s | 0.20 | **1.22** | 1.02 | 0.20 | **0.16** |
|  | 4 | 1.70 | **3.10** | 1.40 | 1.09 | **0.78** |
|  | 10 | 7.83 | 8.45 | **0.62** | 1.44 | 1.21 |
| IPLP | 1 13m16s | 0 | **1.04** | 1.04 | 0 | **−0.02** |
|  | 4 | 0.61 | **2.35** | 1.74 | 0 | **0.04** |
|  | 6 | 4.19 | 5.10 | **0.90** | 0.23 | 0.41 |
| LP | 1 4m23s | 0 | **1.01** | 1.01 | 0 | **−0.05** |
|  | 4 | 1.02 | **2.54** | 1.52 | 0.41 | **0.23** |
|  | 6 | 5.43 | 6.07 | **0.65** | 1.47 | 1.39 |
| Logistic | 1 12m | 0 | **1.14** | 1.14 | 0 | **0.08** |
|  | 4 | 1.28 | **2.64** | 1.36 | 0.669 | **0.33** |
|  | 10 | 7.40 | 7.51 | **0.11** | 1.008 | 0.27 |
| LDF | 1 15m | 11.61 | **12.58** | 0.97 | 11.62 | **11.53** |
|  | 4 | 11.96 | **12.56** | 0.60 | 11.35 | **10.25** |
|  | 6 | 26.87 | 26.72 | **−0.15** | 22.911 | 22.03 |

## 5.4.3 95 % CI of Discriminant Coefficient

Table 5.7 lists the median and 95 % CI of six MP-based LDFs. The three *T*2 coefficients RIP, IPLP, and LP are zeroes, and other coefficients are negative. If four medians of the full model are −0.0128, this LDF is the same as a trivial LDF, such as $f = T1 + T2 + T3 + T4 − 78$ (or 77.5). All MP-based LDFs are almost equal to the trivial LDF in Eq. (5.4).

$$\begin{aligned}F &= -0.0128 \times T1 - 0.0128 \times T2 - 0.0128 \times T3 - 0.0128 \times T4 + 1 \\ &= -T1 - T2 - T3 - T4 + 78\end{aligned} \quad (5.4)$$

Equation (5.5) shows that logistic regression is almost the same as trivial LDF. On the other hand, Fisher's LDF in Eq. (5.6) is different from the trivial LDF.

## 5.4 Pass/Fail Determination by Examination Scores (90 % Level in 2012)

**Table 5.7** 95 % CI of six LDFs

|  |  | T1 | T2 | T3 | T4 | c |
|---|---|---|---|---|---|---|
| RIP | 97.5 % | −0.0075 | 0.0045 | −0.0063 | −0.0048 | 1 |
|  | Median | **−0.0128** | −0.0141 | −0.0150 | **−0.0128** | 1 |
|  | 2.5 % | −0.0190 | −0.0357 | −0.0255 | −0.0147 | 1 |
| H-SVM | 97.5 % | −0.0080 | −0.0020 | −0.0039 | −0.0079 | 1 |
|  | Median | −0.0130 | −0.0130 | −0.0152 | **−0.0128** | 1 |
|  | 2.5 % | −0.0190 | −0.0292 | −0.0228 | −0.0171 | 1 |
| SVM4 | 97.5 % | −0.0080 | −0.0018 | −0.0039 | −0.0079 | 1 |
|  | Median | −0.0129 | −0.0130 | −0.0142 | **−0.0128** | 1 |
|  | 2.5 % | −0.0190 | −0.0292 | −0.0228 | −0.0171 | 1 |
| SVM1 | 97.5 % | −0.0081 | −0.0022 | −0.0039 | −0.0102 | 1 |
|  | Median | −0.0130 | −0.0130 | **−0.0128** | −0.0130 | 1 |
|  | 2.5 % | −0.0188 | −0.0253 | −0.0210 | −0.0171 | 1 |
| IPLP | 97.5 % | −0.0080 | 0.0058 | −0.0062 | −0.004 | 1 |
|  | Median | −0.0130 | −0.0141 | −0.0150 | **−0.0128** | 1 |
|  | 2.5 % | −0.0204 | −0.0358 | −0.0225 | −0.0146 | 1 |
| LP | 97.5 % | −0.0080 | 0.0058 | −0.0072 | −0.0040 | 1 |
|  | Median | −0.0130 | −0.0141 | −0.0150 | **−0.0128** | 1 |
|  | 2.5 % | −0.0190 | −0.0358 | −0.0255 | −0.0146 | 1 |

$$\begin{aligned} \text{Logist } 1234 &= -22.8 \times T1 - 26.84 \times T2 - 27.46 \times T3 - 23.94 \times T4 + 1873.27 \\ &\qquad (1123) \qquad (2945) \qquad (1352) \qquad (1167) \quad (89580) \\ &= -0.0122 \times T1 - 0.0143 \times T2 - 0.0147 \times T3 - 0.0128 \times T4 + 1 \end{aligned}$$

(5.5)

$$\begin{aligned} \text{LDF } 1234 &= 0.026 \times T1 - 0.006 \times T2 + 0.080 \times T3 + 0.009 \times T4 - 1.788 \\ &\qquad (0.013^*) \qquad (0.026) \qquad (0.017^*) \qquad (0.0067) \quad (0.24^*) \\ &= -0.015 \times T1 + 0.003 \times T2 - 0.045 \times T3 - 0.005 \times T4 + 1 \end{aligned}$$

(5.6)

## 5.5 Pass/Fail Determination by Examination Scores (10 % Level in 2012)

In this section, we discuss discrimination at the 10 % level. The passing mark is 36 points. Table 5.1 indicates that four models are linearly separable. A trivial LDF is $f = T1 + T2 + T3 + T4 - 36$ (35.5).

## 5.5.1 MNM and Nine NMs

Table 5.8 lists the MNM and nine "Diff1" of seven LDFs and two discriminant functions. Revised IPLP-OLDF is omitted from the table because it is the same as MNM. None of models has cases on the discriminant hyperplane. SVM1, LDF, QDF, and RDA cannot recognize 12 linear separable models among 16 linear separable models.

H-SVM, SVM4, Revised LP-OLDF, Revised IPLP-OLDF, and logistic regression can recognize four linearly separable models. However, most of the seven nonlinearly separable models are worse than RIP.

## 5.5.2 Error Rate Means (M1 and M2)

Table 5.9 lists the results by Method 1. We obtain the following outcomes: Revised IP-OLDF, Revised IPLP-OLDF, Revised LP-OLDF, and Fisher's LDF select the fourth model as the best model. However, H-SVM, SVM4, SVM1, and logistic regression select the full model as the best model. Because the full model for logistic regression has the minimum $M2$ among all models, we select this as the best model. The seven "$M2$Diff" are −0.07, −0.07, −0.07, 0.12, 0.08, −0.11, and 9.66 %. $M2$ of Revised IP-OLDF is within 0.11 % larger than the four $M2$s of H-SVM, SVM4, SVM1, and logistic regression. Fisher's LDF is 9.66 % larger than Revised IP-OLDF. Among those seven models that are not linearly separable, seven LDFs select fifth models. Moreover, $M2$ of Revised IP-OLDF is the minimum value. The six "$M2$Diff" are 0.4, 0.71, 0, 0.4, 0.39, and 9.6 %. Revised IPLP-OLDF and Revised IP-OLDF are better than the other five LDFs. Fisher's LDF is 9.6 % larger than Revised IP-OLDF. This fact might imply that Revised IP-OLDF is superior to other LDFs for non linearly separable models, not LSD.

**Table 5.8** MNM and nine "Diff1" (10 % level)

| SN | Model | RIP | H-SVM | SVM4 | SVM1 | LP | IPLP | Logistic | LDF | QDF | RDA |
|---|---|---|---|---|---|---|---|---|---|---|---|
| 1 | 1–4 | 0 | 0 | 0 | 0 | 0 | 0 | 0 | 1 | 0 | 1 |
| 2 | 1, 2, 4 | 0 | 0 | 0 | 2 | 0 | 0 | 0 | 1 | 1 | 2 |
| 3 | 2–4 | 0 | 0 | 0 | 0 | 0 | 0 | 0 | 0 | 1 | 2 |
| 4 | 1, 3, 4 | 0 | 0 | 0 | 3 | 0 | 0 | 0 | 1 | 1 | 1 |
| 5 | 1–3 | 2 |  | 1 | 2 | 2 | 0 | 0 | 4 | 0 | 1 |
| 6 | 2, 4 | 4 |  | 2 | 2 | 2 | 0 | 2 | 6 | 2 | 4 |
| 7 | 1, 4 | 3 |  | 3 | 3 | 3 | 0 | 3 | 3 | 3 | 4 |
| 8 | 3, 4 | 3 |  | 1 | 1 | 1 | 0 | 0 | 2 | 2 | 2 |
| 9 | 1, 2 | 7 |  | 0 | 0 | 0 | 0 | 5 | 3 | 0 | 0 |
| 10 | 2, 3 | 7 |  | 4 | 4 | 4 | 0 | 4 | 4 | 4 | 4 |
| 11 | 1, 3 | 8 |  | 2 | 2 | 2 | 0 | 7 | 2 | 3 | 2 |

## 5.5 Pass/Fail Determination by Examination Scores (10 % Level in 2012)

**Table 5.9** *M*1s and *M*2s of eight LDFs (10 %)

| RIP | | *M*1 | *M*2 | Diff. | Model | |
|---|---|---|---|---|---|---|
| 14m8s | 1 | 0 | 0.88 | 0.88 | 1, 2, 3, 4 | |
| | 4 | 0 | **0.86** | 0.86 | 2, 4 | |
| | 5 | 0.73 | **2.44** | 1.71 | 1, 3, 4 | |
| H-SVM | | *M*1 | *M*2 | Diff. | M1Diff. | M2Diff. |
| 5m23s | 1 | 0.00 | **0.81** | 0.81 | 0 | −0.07 |
| | 4 | 0.00 | 0.90 | 0.90 | 0 | 0.03 |
| SVM4 | | *M*1 | *M*2 | Diff. | M1Diff. | M2Diff. |
| 9m23s | 1 | 0 | **0.81** | 0.81 | 0 | −0.07 |
| | 4 | 0 | 0.90 | 0.90 | 0 | 0.04 |
| | 5 | 1.36 | **2.84** | 1.48 | 0.629 | **0.40** |
| SVM1 | | *M*1 | *M*2 | Diff. | M1Diff. | M2Diff. |
| 8m53s | 1 | 0 | **0.81** | 0.81 | 0 | −0.07 |
| | 4 | 0.76 | 1.71 | 0.95 | 0.758 | 0.85 |
| | 5 | 1.65 | **3.15** | 1.50 | 0.919 | **0.71** |
| IPLP | | *M*1 | *M*2 | Diff. | M1Diff. | M2Diff. |
| 13m16s | 1 | 0 | 1.00 | 1.00 | 0 | **0.12** |
| | 4 | 0.00 | **0.85** | 0.85 | 0 | −0.02 |
| | 5 | 0.73 | **2.44** | 1.70 | 0 | **0** |
| LP | | *M*1 | *M*2 | Diff. | M1Diff. | M2Diff. |
| 4m23s | 1 | 0 | **0.96** | 0.96 | 0 | **0.08** |
| | 4 | 0 | **0.80** | 0.80 | 0 | −0.07 |
| | 5 | 1.36 | **2.85** | 1.48 | 0.629 | **0.40** |
| Logistic | | *M*1 | *M*2 | Diff. | M1Diff. | M2Diff. |
| 12m | 1 | 0 | **0.77** | 0.77 | 0 | −0.11 |
| | 4 | 0 | 0.91 | 0.91 | 0 | 0.05 |
| | 5 | 1.59 | **2.83** | 1.24 | 0.855 | **0.39** |
| LDF | | *M*1 | *M*2 | Diff. | M1Diff. | M2Diff. |
| 15m | 1 | 9.64 | 10.54 | 0.90 | 9.637 | **9.66** |
| | 4 | 9.54 | **9.91** | 0.37 | 9.54 | 9.05 |
| | 5 | 11.44 | **12.04** | 0.60 | 10.71 | **9.60** |

### 5.5.3 95 % CI of Discriminant Coefficients

We examine the coefficients of the full model as the best model. A trivial LDF is $f = T1 + T2 + T3 + T4 - 35.5 = -0.028 \times T1 - 0.028 \times T2 - 0.028 \times T3 - 0.028 \times T4 + 1$. Equation (5.7) is Revised IP-OLDF. Equation (5.8) is a logistic regression. Equation (5.9) is Fisher's LDF. The seven LDFs are not the same as trivial LDF. We cannot explain the reason that the full models for all LDFs are not similar to the trivial LDF. This fact is a new complicated research theme for the near future. However, if we compare two results such as 10 and 90 % levels, the former failed class consists of small sample with the unstable answer patterns, and the latter passing class consists of small sample with the stable answer patters because passing students have high scores.

$$\text{Revised IP} - \text{OLDF} = -0.008 \times T1 - 0.061 \times T2 - 0.032 \times T3 \\ - 0.024 \times T4 + 1 \quad (5.7)$$

$$\begin{aligned}\text{Logistic } 1234 &= -2.6 \times T1 - 21 \times T2 - 6.6 \times T3 - 6.97 \times T4 + 296 \\ & \quad (725) \qquad (2489) \qquad (1248) \qquad (724)(26825) \\ &= -0.009 \times T1 - 0.071 \times T2 - 0.022 \times T3 - 0.024 \times T4 + 1.5 \end{aligned} \quad (5.8)$$

$$\begin{aligned}\text{LDF } 1234 &= 0.006 \times T1 + 0.075 \times T2 + 0.007 \times T3 + 0.023 \times T4 - 1.683 \\ & \quad (0.011) \qquad (0.022^*) \qquad (0.015) \quad (0.006^*) \quad (0.201^*) \\ &= -0.004 \times T1 - 0.045 \times T2 - 0.004 \times T3 - 0.014 \times T4 + 1 \end{aligned} \quad (5.9)$$

## 5.6 Summary

In this chapter, we discussed the Method 1 and the best model. We discriminated the pass/fail determinations at 10, 50, and 90 % levels. We selected the best models for eight LDFs by the "minimum $M2$ standard" method. Two studies by 50 and 90 % selected the same best models because only the full models are linearly separable. We obtained surprising results for the best models of all MP-based LDFs and logistic regression that are almost the same as trivial LDFs. Both Fisher's LDFs are quite different from trivial LDFs. We were able to obtain these results by setting the intercept of MP-based LDFs to one. The absolute values for "$M2$Diff" of the six LDFs, with the exception of Fisher's LDF, were within 0.08 and 0.16 %, respectively. However, those of Fisher's LDF were 4.58 and 11.53 %, respectively. Next, we selected the second best model among ten non linearly separable models by six LDFs. The analysis of 50 % selected the second three-variable model, and all "$M2$Diff" were greater than zero. In particular, that of Fisher's LDF was 3.026 %. The analysis of 90 % selected the fourth three-variable model, and all "$M2$Diff" were greater than 0.04 %. In particular, that of Fisher's LDF was 10.247 %. On the other hand, there were four linearly separable models at the 10 % level. Four LDFs, such as RIP, IPLP, LP, and Fisher's LDF, selected the fourth two-variable model, and four LDFs, such as H-SVM, SVM4, SVM1, and logistic regression, selected the full model. We selected the full model as the best model. Moreover, all LDFs were not the same as trivial LDFs.

Based on the above results at the 10, 50, and 90 % levels, we summarize as follows:

## 5.6 Summary

1. The three M2s for Fisher's best LDF are 9.66, 4.58, and 11.53 % worse than Revised IP-OLDF. Only Fisher's LDFs are fragile for the pass/fail determination by examination scores. Therefore, we are concerned about obtaining the same results with medical diagnoses because both data structures are same.
2. The two best Revised IP-OLDF models and logistic regression are the same as trivial LDFs at the 50 and 90 % levels. However, all LDFs are not the same as trivial LDFs at the 10 % level. We cannot explain the reason theoretically.
3. If we select the second best LDF for non linearly separable models, all LDFs select the same models, and $M2$ of Revised IP-OLDFs has the minimum values. This fact might imply that Revised IP-OLDF is superior to other LDFs for the non linearly separable models, although only Revised IP-OLDF and H-SVM can recognize the linearly separable models.
4. If we discriminate the datasets using 100 items as independent variables, we obtain more drastic results. Because I am afraid to be misunderstood as agitator, I do not discuss the results of 100-item discriminations.

## References

Cox DR (1958) The regression analysis of binary sequences (with discussion). J Roy Stat Soc B 20:215–242

Fisher RA (1936) The use of multiple measurements in taxonomic problems. Ann Eugenics 7:179–188

Fisher RA (1956) Statistical methods and statistical inference. Hafner Publishing Co, New Zealand

Friedman JH (1989) Regularized discriminant analysis. J Am Stat Assoc 84(405):165–175

Goodnight JH (1978) SAS technical report—the sweep operator: its importance in statistical computing—(R100). SAS Institute Inc, UK

Lachenbruch PA, Mickey MR (1968) Estimation of error rates in discriminant analysis. Technometrics 10:1–11

Miyake A, Shinmura S (1976) Error rate of linear discriminant function. In: Dombal FT, Gremy F (ed). North-Holland Publishing Company, The Netherland, pp 435–445

Miyake A, Shinmura S (1979) An algorithm for the optimal linear discriminant functions. Proceedings of the international conference on cybernetics and society, pp 1447–1450

Miyake A, Shinmura S (1980) An algorithm for the optimal linear discriminant function and its application. Jpn Soc Med Electron Biol Eng 18(6):452–454

Nomura Y, Shinmura S (1978) Computer-assisted prognosis of acute myocardial infarction. MEDINFO 77, In: Shires, W (ed) IFIP. North-HollandPublishing Company, The Netherland, pp 517–521

Sall JP (1981) SAS regression applications. SAS Institute Inc., USA. (Shinmura S. translate Japanese version)

Sall JP, Creighton L, Lehman A (2004) JMP start statistics, third edition. SAS Institute Inc., USA. (Shinmura S. edits Japanese version)

Schrage L (1991) LINDO—an optimization modeling systems. The Scientific Press, UK. (Shinmura S. & Takamori, H. translate Japanese version)

Schrage L (2006) Optimization modeling with LINGO. LINDO Systems Inc., USA. (Shinmura S. translates Japanese version)

Shimizu T, Tsunetoshi Y, Kono H, Shinmura S (1975). Classification of subjective symptoms of junior high school Students affected by photochemical air pollution. J Jpn Soc Atmos Environ

9/4:734–741. Translated for NERCLibrary, EPA, from the original Japanese by LEO CANCER Associates, P.O.Box 5187 Redwood City, California 94063, 1975 (TR 76-213)

Shinmura S, Miyake A (1979) Optimal linear discriminant functions and their application. COMPSAC 79:167–172

Shinmura S (1984) Medical data analysis, model, and OR. Oper Res 29(7):415–421

Shinmura S, Iida K, Maruyama C (1987) Estimation of the effectiveness of cancer treatment by SSM using a null hypothesis model. Inf Health Social Care 7(3):263–275. doi:10.3109/14639238709010089

Shinmura S (1998) Optimal linear discriminant functions using mathematical programming. J Jpn Soc Comput Stat 11/2:89–101

Shinmura S, Tarumi T (2000) Evaluation of the optimal linear discriminant functions using integer programming (IP-OLDF) for the normal random data. J Jpn Soc Comput Stat 12(2):107–123

Shinmura S (2000a) A new algorithm of the linear discriminant function using integer programming. New Trends Prob Stat 5:133–142

Shinmura S (2000b) Optimal linear discriminant function using mathematical programming. Dissertation, 200:1–101, Okayama University, Japan

Shinmura S, Suzuki T, Koyama H, Nakanishi K (1983) Standardization of medical data analysis using various discriminant methods on a theme of breast diseases. MEDINFO 83, In: Van Bemmel JH, Ball MJ, Wigertz O (ed). North-Holland Publishing Company, The Netherland, pp 349–352

Shinmura S (2001) Analysis of effect of SSM on 152,989 cancer patient. ISI2001, pp 1–2. doi:10.13140/RG.2.1.30779281

Shinmura S (2003) Enhanced algorithm of IP-OLDF. ISI2003 CD-ROM, pp 428–429

Shinmura S (2004) New algorithm of discriminant analysis using integer programming. IPSI 2004 pescara VIP conference CD-ROM, pp 1–18

Shinmura S (2005) New age of discriminant analysis by IP-OLDF—beyond Fisher's linear discriminant function. ISI2005, pp 1–2

Shinmura S (2007) Overviews of discriminant function by mathematical programming. J Jpn Soc Comput Stat 20(1-2):59–94

Shinmura S (2010a) The optimal linearly discriminant function (Saiteki Senkei Hanbetu Kansuu). Union of Japanese Scientist and Engineer Publishing, Japan

Shinmura S (2010b) Improvement of CPU time of Revised IP-OLDF using linear programming. J Jpn Soc Comput Stat 22(1):39–57

Shinmura S (2011a) Beyond Fisher's linear discriminant analysis—new world of the discriminant analysis. 2011 ISI CD-ROM, pp 1–6

Shinmura S (2011b) Problems of discriminant analysis by mark sense test data. Jpn Soc Appl Stat 40(3):157–172

Shinmura S (2013) Evaluation of optimal linear discriminant function by 100-fold cross-validation. ISI CD-ROM, pp 1–6

Shinmura S (2014a) End of discriminant functions based on variance-covariance matrices. ICORE2014, pp 5–16

Shinmura S (2014b) Improvement of CPU time of linear discriminant functions based on MNM criterion by IP. Stat Optim Inf Comput 2:114–129

Shinmura S (2014c) Comparison of linear discriminant functions by $K$-fold cross-validation. Data Anal 2014:1–6

Shinmura S (2015a) The 95 % confidence intervals of error rates and discriminant coefficients. Stat Optim Inf Comput 2:66–78

Shinmura S (2015b) A trivial linear discriminant function. Stat Optim Inf Comput 3:322–335. doi:10.19139/soic.20151202

Shinmura S (2015c) Four serious problems and new facts of the discriminant analysis. In: Pinson E, Valente F, Vitoriano B (ed) Operations research and enterprise systems, pp 15–30. Springer, Berlin (ISSN: 1865-0929, ISBN: 978-3-319-17508-9, doi:10.1007/978-3-319-17509-6)

Shinmura S (2015d) Four problems of the discriminant analysis. ISI 2015:1–6

# References

Shinmura S (2016a) The best model of Swiss banknote data. Stat Optim Inf Comput 4:118–131. International Academic Press (ISSN: 2310-5070 (online) ISSN: 2311-004X (print), doi: 10.19139/soic.v4i2.178)

Shinmura S (2016b) Matroska feature selection method for microarray data. Biotechno 2016:1–8

Shinmura S (2016c) Discriminant analysis of the linear separable data—Japanese automobiles. J Stat Sci Appl X, X:0–14

VapnikV (1995) The nature of statistical learning theory. Springer, Berlin

# Chapter 6
# Best Model for Swiss Banknote Data

## Explanation 1 of Matroska Feature-Selection Method (Method 2)

## 6.1 Introduction

In this chapter, we discuss Problem 4 in addition to Problems 2 and 5, where the discriminant analysis is not the inferential statistical method (Shinmura 2014a, 2015c, d). We propose a *k*-fold cross-validation for small sample method (Method 1) and can obtain the 95 % CI of error rates (Miyake and Shinmura 1976) and discriminant coefficients. Through this innovation, we can select the best model with the minimum error rate mean in the validation samples (Minimum M2 Standard) (Shinmura 2014c, 2016c). We examine this new model selection procedure instead of leave-one-out (LOO) procedure (Lachenbruch and Mickey 1968) through many data and obtain excellent results. However, we cannot explain the useful meaning of the 95 % CI of discriminant coefficients (Shinmura 2010a). After many trials, we set the intercept to one for eight linear discriminant functions (LDFs). Six MP-based LDFs are revised optimal linear discriminant function using integer programming (Revised IP-OLDF) based on minimum number of misclassifications (MNM) (Miyake and Shinmura 1979; Shinmura 2011a, b, 2013), Revised LP-OLDF, Revised IPLP-OLDF (Shinmura 2010b, 2014b), three support vector machines (SVMs) (Vapnik 1995). Two statistical LDFs are logistic regression (Cox 1958, Firth 1993), and Fisher's LDF (Fisher 1936, 1956). Seven LDFs, with the exception of Fisher's LDF, are almost the same as a trivial LDF for six pass/fail determinations that use examination scores (Shinmura 2015a, b), the full model of those is linearly separable. In this chapter, we examine the 16 linearly separable models of Swiss banknote data (Flury and Rieduyl 1988) by eight LDFs. *M2* of the best model of Revised IP-OLDF is the smallest value of all models. We find all coefficients of the best model rejected by the 95 % CI of discriminant coefficients (Coefficient Standard). We compare *t*-values of the discriminant scores instead of *p*-values because the range of *p*-values is [0, 1] and narrow. The *t*-value of the best model has the maximum values among 16 models (maximum *t*-value Standard). Therefore, both standards support the best model of Revised IP-OLDF in these data.

Moreover, we study LSD discrimination through these data in addition to Japanese-automobile data (Shinmura 2016c) and six pass/fail determinations using examination scores precisely.

We propose the Matroska feature-selection method for microarray dataset (Method 2) (Shinmura 2015e–s, 2016b) in Chap. 8. Because LSD discrimination is no longer popular, we explain this new Method 2 through detailed examples of Swiss banknote and Japanese-automobile data. In the gene analysis, we call all linearly separable models, "Matroska." The full model is the largest Matroska that includes all smaller Matroskas in it. We already know that the smallest Matroska (basic gene set, BGS) can explain the Matroska structure completely by the monotonic decrease of MNM (MNM$_p$ ≥ MNM$_{(p+1)}$). On the other hand, LASSO (Buhlmann and Geer 2011; Simon et al. 2013) attempts to make feature-selection. If it cannot find the smallest Matroska in the data, it cannot explain the Matroska structure. Swiss banknote data, Japanese-automobile data, and six microarray datasets (Jeffery et al. 2006) are helpful for evaluating the usefulness of other feature-selection methods, including LASSO.

## 6.2 Swiss Banknote Data

In this chapter, we discriminate Swiss banknote data and its resampling samples by eight LDFs. We focus on two error rates means, *M*1 and *M*2, from the training and validation samples, respectively, and propose the model with minimum *M*2 as the best model instead of the LOO procedure. We compare eight *M*2s of the best model to eight LDFs and determine the best model among the eight LDFs. Moreover, we discuss the discriminant coefficients by setting the intercept to one. We confirm the validity of the best model by both the coefficient standard and maximum *t*-value standard by the t test of two means.

### 6.2.1 Data Outlook

Swiss banknote data consist of two types of bills, such as 100 genuine ($y_i = 1$) and 100 counterfeit ($y_i = -1$) bills. There are six variables, such as *X*1, which is the bill length; *X*2 and *X*3, which are the widths of the left and right edges, respectively; *X*4 and *X*5, which are the bottom and top margin widths, respectively; and *X*6, which is the length of the image diagonal. We can download these data from the Internet.

Table 6.1 lists the full model by regression analysis (plug-in rule1). We determine that only the *X*1 coefficient is accepted at the 5 % level by *p*-value. Forward stepwise selects the variables as follows: *X*6, *X*4, *X*5, *X*3, *X*2, and *X*1. We select the models with minimum values of AIC, BIC, and |Cp − (*p* + 1) |. AIC selects the five-variable model (*X*2–*X*6), BIC selects the three-variable model (*X*4–*X*6), and Cp selects the full model. Because of the AIC, BIC, and Cp statistics select three

## 6.2 Swiss Banknote Data

**Table 6.1** Full model by regression analysis (*left*) and four statistics by forward stepwise technique (*right*)[*]

| Var. | Regression analysis |      |        |      | Forward stepwise technique |         |         |        |
|------|---------------------|------|--------|------|------|---------|---------|--------|
|      | Coeff.              | SE   | t      | p    | $R^2$ | AIC     | BIC     | Cp     |
| c    | 24.09               | 6.55 | 3.68   | 0.00 | –    | –       | –       | –      |
| X6   | −0.21               | 0.02 | −13.90 | 0.00 | 0.81 | −34.16  | −24.39  | 292.02 |
| X4   | 0.15                | 0.01 | 14.77  | 0.00 | 0.88 | −128.72 | −115.74 | 107.00 |
| X5   | 0.16                | 0.02 | 9.22   | 0.00 | 0.92 | 205.74  | **189.56** | 10.66 |
| X3   | 0.11                | 0.04 | 2.77   | 0.01 | 0.92 | 205.99  | 186.64  | 10.26  |
| X2   | 0.12                | 0.04 | 2.69   | 0.01 | 0.92 | **210.90** | 188.40 | 5.32 |
| X1   | 0.02                | 0.03 | 0.56   | **0.57** | 0.92 | −209.06 | −183.43 | **7.00** |

[*] We computed this table again in September 2015. Some values are different from the old table

different models, and thus, we cannot determine the proper model uniquely. On the other hand, IP-OLDF finds that the two-variable model (X4, X6) is linearly separable through examination of all the possible combinations of six independent variables (Goodnight 1978). Two-variable model (X4, X6) is BGS and can explain the structure of Swiss banknote data that is very important for us to understand Method 2 in Chap. 8. Because of the monotonic decrease of MNM ($MNM_p \geq MNM_{(p+1)}$), we know that the 16 models, including (X4, X6), are linearly separable. On the other hand, other 47 models are not linearly separable. Therefore, we can select the best model among these 16 linearly separable models. In the gene analysis, we call the linearly separable model, "Matroska." The full model is the largest Matroska that includes all smaller Matroska in it. We call the smallest Matroska, such as (X4, X6), "the basic gene set (BGS)." We can explain the Matroska structure by BGS completely.

Figure 6.1 is a scatter plot by (X4, X6). Genuine and counterfeit bills are represented by the symbols "O" and "×," respectively. The two circles are 99 % confidence probability ellipses that are expected to include 99 % bills in each ellipse if the two classes are supposed to be normal distributions. We understand that

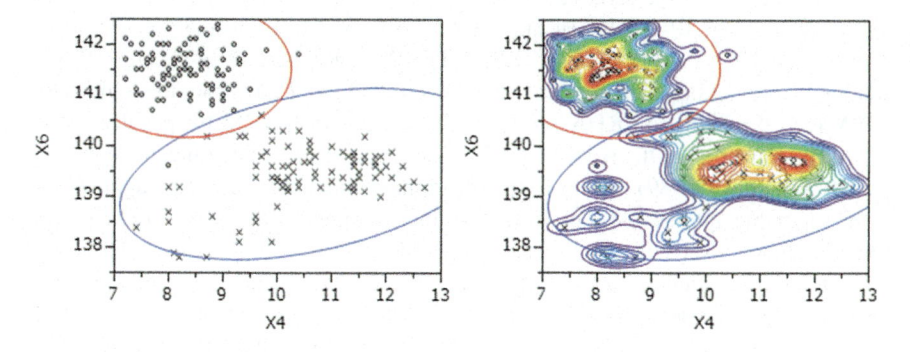

**Fig. 6.1** Swiss banknote data ($MNM_{(X4,X6)} = 0$)

counterfeit bills are not well controlled because their variance is significant. There might reasons for not finding the data to be linearly separable, as follows:

1. To this point, only H-SVM has been able to recognize linearly separable models theoretically. However, it can be adopted only for linearly separable models. Therefore, and to the best of our knowledge, no one has attempted to use H-SVM for discrimination. In addition, SVM researchers are interested in kernel SVM because its idea is attractive.
2. To the best of our knowledge, nobody has considered the importance of LSD discrimination. Many scientists have claimed that the purpose of discriminant analysis is to discriminate overlapping data, not LSD. However, all LDFs, with the exception of H-SVM and Revised IP-OLDF, cannot discriminate overlapping data correctly because "MNM = 0" indicates that the data do not overlap.
3. All possible combinations of regression models (Goodnight 1978; Sall 1981) provide clear and deterministic data perception. Therefore, we discriminate as many possible combinations of discriminant models as possible. If some researchers attempt to verify all scatter plot combinations, as shown in Fig. 6.1, they might suspect that the model (X4, X6) is linearly separable. IP-OLDF (Shinmura 1998, 2000a, b, 2003, 2004, 2005, 2007; Shinmura and Tarumi 2000) finds that these data are linearly separable through examination of 63 discriminations.

## 6.2.2 Comparison of Seven LDF for Original Data

In this chapter, we investigate a total of 63 ($=2^6 - 1$) discriminant models. A total of 16 models, including the model (X4, X6), are linearly separable. Other 47 models are not linearly separable. These data are adequate regardless of whether eight LDFs can discriminate linearly separable models correctly. We focus on the linearly separable models because evaluation is very explicit. Table 6.2 lists the results of the 16 linearly separable models. In the table, "SN" is the sequential number of the discriminant model; "$p$" is the number of variables; "Var" is a suffix of the variable; "1–6" indicates the six-variable model (X1, X2, X3, X4, X5, X6); and "RIP" is MNM of Revised IP-OLDF. Because NMs of H-SVM, a soft-margin SVM for penalty $c = 10^4$ (SVM4), Revised LP-OLDF (LP), Revised IPLP-OLDF (IPLP), and logistic regression (logistic) are zero, we omitted five LDFs from the table. The "SVM1, LDF, QDF, and RDA" columns show NMs of S-SVM for penalty $c = 1$ (SVM1), Fisher's LDF (LDF), QDF, and regularized discriminant analysis (RDA) (Friedman 1989). These four discriminant functions cannot recognize linearly separable models. We observe this fact in other data. Therefore, we can conclude that SVM1 with small penalty $c$, Fisher's LDF, QDF, and RDA based on the variance–covariance matrices are weak for the discrimination of linearly separable models (Problem 2). None of the LDFs, with the exception of Revised IP-OLDF, can discriminate cases on the discriminant hyperplane theoretically. We cannot determine whether the four statistical discriminant functions do not verify

## 6.2 Swiss Banknote Data

**Table 6.2** MNM and NMs of 16 linearly separable models

| SN | p | var. | RIP | SVM1 | LDF | QDF | RDA |
|---|---|---|---|---|---|---|---|
| 1 | 6 | 1–6 | 0 | 1 | 1 | 1 | 1 |
| 2 | 5 | 2–6 | 0 | 1 | 1 | 1 | 1 |
| 3 | 5 | 1, 3–6 | 0 | 1 | 1 | 1 | 1 |
| 4 | 5 | 1, 2, 4–6 | 0 | 1 | 1 | 1 | 1 |
| 5 | 5 | 1–4, 6 | 0 | 2 | 1 | 1 | 1 |
| 6 | 4 | 3–6 | 0 | 1 | 1 | 1 | 1 |
| 7 | 4 | 2, 4–6 | 0 | 1 | 1 | 1 | 1 |
| 8 | 4 | 1, 4–6 | 0 | 1 | 1 | 1 | 1 |
| 9 | 4 | 2–4, 6 | 0 | 2 | 1 | 1 | 1 |
| 10 | 4 | 1, 3, 4, 6 | 0 | 2 | 1 | 1 | 1 |
| 11 | 4 | 1, 2, 4, 6 | 0 | 2 | 2 | 1 | 1 |
| 12 | 3 | 4–6 | 0 | 1 | 1 | 1 | 1 |
| 13 | 3 | 3, 4, 6 | 0 | 2 | 1 | 1 | 1 |
| 14 | 3 | 1, 4, 6 | 0 | 2 | 2 | 2 | 1 |
| 15 | 3 | 2, 4, 6 | 0 | 2 | 1 | 1 | 1 |
| 16 | 2 | 4, 6 | 0 | 2 | 3 | 1 | 1 |

the number of cases on the discriminant hyperplane (Problem 1). LINGO solves the six MP-based LDFs (Schrage 1991, 2006), and JMP (Sall et al. 2004) solves the statistical discriminant function such as Fisher's LDF, logistic regression, QDF, and RDA. Although Fisher's LDF and logistic regression are LDFs, QDF and RDA are not LDFs. Therefore, we do not evaluate QDF and RDA by Method 1.

Table 6.3 lists NMs of 11 models that exclude $X6$. The bold NMs are the same as MNM. We can understand that MNMs are the minimum NMs of nine discriminant functions, with the exception of H-SVM. Revised IPLP-OLDF is the second best because it obtains the MNM estimate, and ten NMs are the same as MNM. Next, we observe that logistic regression is often better than the other six LDFs.

**Table 6.3** MNM and eight NMs of 11 nonlinearly separable models

| SN | Var. | RIP | SVM4 | SVM1 | LP | IPLP | logistic | LDF | QDF | RDA |
|---|---|---|---|---|---|---|---|---|---|---|
| 17 | 1–5 | **2** | 3 | 3 | 3 | **2** | **2** | 7 | 6 | 6 |
| 18 | 1, 3–5 | **2** | 3 | 3 | 3 | **2** | **2** | 7 | 6 | 6 |
| 19 | 2–5 | **2** | 3 | 5 | 3 | **2** | **2** | 6 | 6 | 5 |
| 20 | 1, 2, 4, 5 | **2** | 3 | 5 | 3 | **2** | **2** | 8 | 6 | 8 |
| 21 | 4 | **12** | 17 | 17 | 17 | **12** | 15 | 19 | 14 | 14 |
| 22 | 3–5 | **2** | 3 | 4 | 3 | **2** | **2** | 6 | 6 | 5 |
| 23 | 2, 4, 5 | **2** | **2** | 5 | **2** | **2** | **2** | 8 | 6 | 5 |
| 24 | 1, 4, 5 | **2** | 4 | 5 | 4 | **2** | **2** | 9 | 6 | 6 |
| 25 | 1, 3, 4 | **13** | 17 | 17 | 17 | 14 | 15 | 19 | 16 | 16 |
| 26 | 2–4 | **13** | 17 | 17 | 17 | **13** | 17 | 19 | 15 | 15 |
| 27 | 1, 2, 4 | **13** | 19 | 19 | 19 | **13** | 16 | 22 | 18 | 18 |

**Table 6.4** Comparison of eight discriminant functions

| | $c$ | $b$ | $R^2$ | MNM = 1 | MNM = 40 |
|---|---|---|---|---|---|
| SVM4 | 0.72 | 1.04 | 0.988 | 1.76 | 42.32 |
| SVM1 | 1.57 | 1.02 | 0.989 | **2.59** | 42.37 |
| LP | −0.03 | 0.97 | 0.998 | 0.94 | 38.77 |
| IPLP | 0.68 | 1.03 | 0.987 | 1.71 | 41.88 |
| Logistic | 0.32 | 1 | 0.995 | 1.32 | 40.32 |
| LDF | 1.28 | 1.11 | 0.974 | 2.38 | **45.68** |
| QDF | 1.59 | 0.99 | 0.991 | **2.58** | 41.19 |
| RDA | 1.56 | 0.99 | 0.992 | **2.55** | 41.16 |

Because we had no validation samples before Method 1, we evaluated different LDFs in the training samples using simple regression as explained here (Shinmura 2010a). We could evaluate all LDFs by MNM in the training samples because MNM is the minimum NMs among all LDFs. In our research, we also evaluate QDF and RDA by MNM. These are not LDFs. Table 6.4 lists the results of regression analysis, such as "each NM = $c + b \times$ MNM" using 63 NMs, including 16 linearly separable models. QDF is the worst result because the intercept is 1.59, and the discriminant coefficient is 0.99, which is almost 1. The simple regression line (QDF = 1.59 + 0.99 × MNM) can predict good NMs of QDF by MNMs because $R$-square is 0.991. This result implies that NMs of QDF are almost 1.59 higher than MNMs. In addition, the error rate of QDF is 0.8 % (=1.59/200) greater than that of Revised IP-OLDF. We conclude that NMs of QDF, RDA, LDF, and SVM1 are at least 1.28 higher than MNMs. On the contrary, NMs of SVM4, Revised LP-OLDF, Revised IPLP-OLDF, and logistic regression are half of NMs for these discriminant functions. Revised IPLP-OLDF is expected to be a good estimate of MNM. However, Revised LP-OLDF and logistic regression are better than Revised IPLP-OLDF for 63 models. We agree that everyone might not accept this explanation on the superiority of Revised IP-OLDF.

## 6.3 100-Fold Cross-Validation for Small Sample Method

We generate resampling samples from Swiss banknote data and evaluate eight LDFs by our new Method 1.

### 6.3.1 Best Model Comparison

Table 6.5 lists the 16 linearly separable models and 23th model, that is not linearly separable model, in Table 6.3. "$M1$ and $M2$" are the mean error rates in the training and validation samples. All 16 $M$1s of Revised IP-OLDF, H-SVM, SVM4, LP,

## 6.3 100-Fold Cross-Validation for Small Sample Method

**Table 6.5** 100-fold cross-validation for small sample method

| LDF | SN | M1 | M2 | t | Diff. | Model |
|---|---|---|---|---|---|---|
| RIP 53m42s | 1 | 0 | **0.30** | **453** | 0.30 | 1–6 |
| | 2 | 0 | 0.77 | 307 | 0.77 | 2–6 |
| | 3 | 0 | **0.26** | **456** | 0.26 | 1, 3–6 |
| | 4 | 0 | **0.30** | **453** | 0.30 | 1, 2, 4–6 |
| | 5 | 0 | 0.70 | 243 | 0.70 | 1–4, 6 |
| | 6 | 0 | 0.74 | 409 | 0.74 | 3–6 |
| | 7 | 0 | 0.75 | 419 | 0.75 | 2, 4–6 |
| | 8 | 0 | **0.27** | **454** | 0.27 | 1, 4–6 |
| | 9 | 0 | 0.77 | 362 | 0.77 | 2–4, 6 |
| | 10 | 0 | 0.63 | 379 | 0.63 | 1, 3, 4, 6 |
| | 11 | 0 | 0.62 | 379 | 0.62 | 1, 2, 4, 6 |
| | 12 | 0 | **0.69** | **402** | 0.69 | 4–6 |
| | 13 | 0 | 0.67 | 353 | 0.67 | 3, 4, 6 |
| | 14 | 0 | 0.60 | 379 | 0.60 | 1, 4, 6 |
| | 15 | 0 | 0.66 | 366 | 0.66 | 2, 4, 6 |
| | 16 | 0 | **0.47** | **359** | 0.47 | 4, 6 |
| | 23 | 0.84 | **1.69** | 315 | 0.85 | 2, 4, 5 |

| | SN | M1 | M2 | t | Diff | M1Diff. | M2Diff. |
|---|---|---|---|---|---|---|---|
| H-SVM 35m6s | 1 | 0 | 0.53 | −147 | 0.53 | 0.00 | 0.23 |
| | 2 | 0 | 0.46 | 182 | 0.46 | 0.00 | −0.30 |
| | 3 | 0 | 0.46 | −163 | 0.46 | 0.00 | **0.21** |
| | 4 | 0 | 0.45 | −158 | 0.45 | 0.00 | 0.15 |
| | 5 | 0 | 0.72 | 141 | 0.72 | 0.00 | 0.02 |
| | 6 | 0 | 0.46 | **192** | 0.46 | 0.00 | −0.28 |
| | 7 | 0 | 0.43 | −185 | 0.43 | 0.00 | −0.32 |
| | 8 | 0 | **0.38** | **−164** | 0.38 | 0.00 | **0.11** |
| | 9 | 0 | 0.70 | 149 | 0.70 | 0.00 | −0.06 |
| | 10 | 0 | 0.66 | 147 | 0.66 | 0.00 | 0.03 |
| | 11 | 0 | 0.65 | 143 | 0.65 | 0.00 | 0.03 |
| | 12 | 0 | **0.39** | **184** | 0.39 | 0.00 | −0.30 |
| | 13 | 0 | 0.63 | 147 | 0.63 | 0.00 | −0.04 |
| | 14 | 0 | 0.60 | 142 | 0.60 | 0.00 | −0.01 |
| | 15 | 0 | 0.59 | 142 | 0.59 | 0.00 | −0.07 |
| | 16 | 0 | **0.46** | **140** | 0.46 | 0.00 | −0.01 |
| SVM4 44m46s | 3 | 0 | 0.464 | | 0.46 | 0.00 | **0.21** |
| | 8 | 0 | **0.374** | | 0.37 | 0.00 | **0.10** |
| | 23 | 1.21 | **1.764** | | 0.56 | 0.37 | **0.08** |
| SVM1 46m17s | 3 | 0.26 | 0.54 | | 0.28 | 0.26 | **0.28** |
| | 12 | 0.32 | **0.52** | | 0.21 | 0.32 | −0.17 |
| | 23 | 1.94 | **2.52** | | 0.58 | 1.11 | **0.84** |

(continued)

**Table 6.5** (continued)

|  | SN | M1 | M2 | t | Diff | M1Diff. | M2Diff. |
|---|---|---|---|---|---|---|---|
| IPLP 47m31s | 3 | 0 | 0.49 |  | 0.49 | 0.00 | **0.23** |
|  | 8 | 0 | **0.41** |  | 0.41 | 0.00 | **0.14** |
|  | 23 | 0.84 | **1.66** |  | 0.82 | 0.00 | -0.03 |
| LP 19m58s | 3 | 0.00 | 0.27 |  | 0.27 | 0.00 | **0.01** |
|  | 8 | 0.00 | **0.27** |  | 0.27 | 0.00 | **0.00** |
|  | 23 | 1.22 | **1.81** |  | 0.59 | 0.38 | **0.12** |
| Logistic 46m | 3 | 0.00 | 0.52 |  | 0.52 | 0.00 | **0.26** |
|  | 12 | 0.00 | **0.41** |  | 0.41 | 0.00 | **−0.27** |
|  | 23 | 1.51 | **2.02** |  | 0.51 | 0.67 | **0.33** |
| LDF 55m | 3 | 0.53 | 0.55 |  | 0.02 | 0.53 | **0.29** |
|  | 7 | 0.51 | **0.54** |  | 0.03 | 0.51 | **−0.20** |
|  | 23 | 3.10 | **3.43** |  | 0.33 | 2.27 | **1.75** |

IPLP, and logistic regression are zero. SVM1 and Fisher's LDF cannot recognize all linearly separable models. Only Revised IP-OLDF selects the third five-variable model as the best model, with an M2 of 0.26 %. It's t value of two class means on discriminant score is maximum value 456 among 17 models. H-SVM, SVM4, Revised IPLP-OLDF, and Revised LP-OLDF select the eighth four-variable model as the best model, with M2s of 0.38, 0.37, 0.41, and 0.27 %, respectively. SVM1 and logistic regression select the 12th three-variable model, with M2s of 0.52 and 0.41 %, respectively. Only Fisher's LDF selects the seventh four-variable model, with M2 of 0.54 %. The best model of Revised IP-OLDF has the minimum value of M2 among eight LDFs. The seven "M2Diff" of third model (X1, X3, X4, X5, X6) are 0.21, 0.21, 0.28, 0.23, 0.01, 0.26, and 0.29 %. Next, we examine the best model among 47 models that are not linearly separable models. We focus on the 23rd three-variable model (X2, X4, X5) of Revised IPLP-OLDF as the best model. Because six "M2Diffs" of (X2, X4, X5) are 0.08, 0.84, −0.03, 0.12, 0.33, and 1.75 %, Revised IPLP-OLDF and Revised IP-OLDF are better than other five LDFs among 47 models.

## 6.3.2 95 % CI of Discriminant Coefficient

### 6.3.2.1 Consideration of 27 Models

We can obtain 95 % CI of the coefficient with our new Method 1. We use the median as eight LDFs. Table 6.6 lists the results of the median represented by a given symbol. In the table, "SN" is the sequential number of 27 models. First, 16

## 6.3 100-Fold Cross-Validation for Small Sample Method

**Table 6.6** 95 % CI of 27 models (First rows of LDF and logistic regression are NMs of two LDFs).

| SN | Model | RIP | LP | IPLP | H-SVM | SVM4 | SVM1 | LDF | Logistic |
|---|---|---|---|---|---|---|---|---|---|
| 1 | 1–6 | +ZZ++−Z | 0Z0- -+Z | 000000* | 0000001 | 0000001 | 0000001 | 1 | 0 |
|   |   | +ZZ++−Z | 000- -+0 | 0Z0- -+Z | 000++−0 | 000++−0 | 000++−0 | 0+ - - -+- | 0000000 |
| 2 | 2–6 | $d$0000* | $d$00000* | $d$00000* | $d$000001 | $d$000000 | $d$000001 | 1 | 0 |
|   |   | $d$00++−0 | $d$00- -+0 | $d$00- -+0 | $d$00++−* | $d$00++−0 | $d$00++−0 | $d$+- - -+- | $d$000000 |
| 3 | 1, 3–6 | +$dZ$++−Z | −$d$0- -+Z | 0$d$0000* | 0$d$00001 | 0$d$00001 | 0$d$00001 | 1 | 0 |
|   |   | +$dZ$++−Z | 0$d$0- -+0 | 0$d$0- -+0 | 0$d$0++−0 | 0$d$0++−0 | 0$d$0++−0 | 0$d$0++−* | 0$d$00000 |
| 4 | 1, 2, 4–6 | +$Zd$++−Z | 00$d$- -+* | 00$d$000* | 00$d$0001 | 00$d$0001 | 00$d$0001 | 1 | 0 |
|   |   | +$Zd$++−Z | 00$d$- -+0 | 00$d$- -++ | 00$d$++−0 | 00$d$++−0 | 00$d$++−0 | 00$d$++−+ | 00$d$0000 |
| 5 | 1–4, 6 | 000+$d$−* | 0000$d$0* | 0000$d$0* | 0000$d$01 | 0000$d$01 | 0000$d$01 | 1 | 0 |
|   |   | 000+$d$−+ | 000- -$d$+0 | 0000$d$+0 | 000+$d$−0 | 000+$d$−0 | 000+$d$−0 | 0−++$d$−+ | 0000$d$00 |
| 6 | 3–6 | $dd$0000* | $dd$0000* | $dd$0000* | $dd$0++−1 | $dd$0++−1 | $dd$0++−1 | 1 | 0 |
|   |   | $dd$0++−0 | $dd$0- -+0 | $dd$0- -+0 | $dd$0++−+ | $dd$0++−0 | $dd$0++−0 | $dd$0- -+− | $dd$00000 |
| 7 | 2, 4–6 | $d$0$d$000* | $d$0$d$000* | $d$0$d$000* | $d$0$d$++−1 | $d$0$d$0001 | $d$0$d$++−1 | 1 | 0 |
|   |   | $d$0$d$++−0 | $d$0$d$- -+0 | $d$0$d$- -+0 | $d$0$d$++−+ | $d$0$d$++−0 | $d$0$d$++−0 | $d$0$d$++−* | $d$0$d$0000 |
| 8 | 1, 4–6 | +$dd$++−Z | 0$dd$- -+* | 0$dd$000* | 0$dd$0001 | 0$dd$0001 | 0$dd$0001 | 1 | 0 |
|   |   | +$dd$++−Z | 0$dd$- -+0 | 0$dd$- -++ | 0$dd$++−0 | 0$dd$++−0 | 0$dd$++−0 | 0$dd$++−+ | 0$dd$0000 |
| 9 | 2–4, 6 | $d$000$d$0* | $d$000$d$0* | $d$000$d$01 | $d$00+$d$−1 | $d$000$d$−1 | $d$00+$d$−1 | 1 | 0 |
|   |   | $d$000$d$−0 | $d$00−$d$+− | $d$000$d$+0 | $d$00+$d$−+ | $d$000$d$−+ | $d$000$d$−+ | $d$−++$d$−+ | $d$000$d$00 |
| 10 | 1, 3, 4, 6 | 0$d$00$d$0* | 0$d$00$d$0* | 0$d$00$d$0* | 0$d$00$d$01 | 0$d$00$d$01 | 0$d$00$d$01 | 1 | 0 |
|   |   | 0$d$0+$d$−0 | 0$d$0−$d$+0 | 0$d$00$d$+0 | 0$d$0+$d$−0 | 0$d$0+$d$−0 | 0$d$0+$d$−0 | −$d$++$d$−+ | 0$d$00$d$00 |
| 11 | 1, 2, 4, 6 | 00$d$+$d$−* | 00$d$0$d$0* | 00$d$0$d$0* | 00$d$0$d$01 | 00$d$0$d$01 | 00$d$0$d$01 | 2 | 0 |
|   |   | 00$d$+$d$−0 | 00$d$−$d$+0 | 00$d$−$d$+0 | 00$d$+$d$−0 | 00$d$+$d$−0 | 00$d$+$d$−0 | 00$d$+$d$−+ | 00$d$0$d$00 |
| 12 | 4–6 | $ddd$++−* | $ddd$000* | $ddd$+00* | $ddd$++−1 | $ddd$++−1 | $ddd$++−1 | 1 | 0 |
|   |   | $ddd$++−0 | $ddd$- -+− | $ddd$- -+0 | $ddd$++−+ | $ddd$++−+ | $ddd$++−+ | $ddd$++−+ | $ddd$0000 |
| 13 | 3, 4, 6 | $dd$00$d$−* | $dd$00$d$0* | $dd$00$d$01 | $dd$0+$d$−1 | $dd$00$d$−1 | $dd$0+$d$−1 | 1 | 0 |
|   |   | $dd$0+$d$−+ | $dd$0−$d$+− | $dd$00$d$+− | $dd$0+$d$−+ | $dd$00$d$−+ | $dd$00$d$−+ | $dd$++$d$−+ | $dd$00$d$00 |
| 14 | 1, 4, 6 | 0$dd$+$d$−* | 0$dd$0$d$0* | 0$dd$0$d$0* | 0$dd$+$d$−1 | 0$dd$0$d$01 | 0$dd$+$d$−1 | 2 | 0 |
|   |   | 0$dd$+$d$−+ | 0$dd$−$d$+− | 0$dd$−$d$+0 | 0$dd$+$d$−+ | 0$dd$+$d$−0 | 0$dd$+$d$−0 | 0$dd$+$d$−+ | 0$dd$0$d$00 |
| 15 | 2, 4, 6 | $d$0$d$+$d$−1 | $d$0$d$0$d$01 | $d$0$d$0$d$01 | $d$0$d$+$d$−1 | $d$0$d$+$d$−1 | $d$0$d$+$d$−1 | 1 | 0 |
|   |   | $d$0$d$+$d$−+ | $d$0$d$−$d$+− | $d$0$d$−$d$+0 | $d$0$d$+$d$−+ | $d$0$d$+$d$−+ | $d$0$d$+$d$−+ | $d$0$d$+$d$−+ | $d$0$d$0$d$00 |
| 16 | 4, 6 | $ddd$+$d$−1 | $ddd$+$d$−1 | $ddd$+$d$−1 | $ddd$+$d$−1 | $ddd$+$d$−1 | $ddd$+$d$−1 | 3 | 0 |
|   |   | $ddd$+$d$−+ | $ddd$+$d$+− | $ddd$−$d$+− | $ddd$+$d$−+ | $ddd$+$d$−+ | $ddd$+$d$−+ | $ddd$−$d$+− | $ddd$0$d$00 |
| 22 | 3–5 | $dd$000$d$1 | $dd$000$d$1 | $dd$000$d$1 | $dd$- - -$d$1 | $dd$- - -$d$1 | $dd$- - -$d$1 | 6 | 2 |
|   |   | $dd$0++$d$0 | $dd$- - -$d$+ | $dd$0- -$d$0 | $dd$+++$d$− | $dd$0++$d$0 | $dd$+++$d$− | $dd$0++$d$− | $dd$0- -$d$+ |
| 25 | 1, 3, 4 | 0$d$00$dd$1 | 0$d$00$dd$* | 0$d$00$dd$1 | 0$d$00$dd$1 | 0$d$00$dd$1 | 0$d$00$dd$1 | 19 | 15 |
|   |   | 0$d$0+$dd$0 | +$d$- -$dd$0 | +$d$- -$dd$0 | −$d$++$dd$0 | −$d$++$dd$0 | −$d$++$dd$0 | −$d$++$dd$− | +$d$- -$dd$+ |
| 27 | 1, 2, 4 | 00$d$0$dd$1 | 00$d$0$dd$* | 00$d$0$dd$1 | 00$d$0$dd$1 | 00$d$0$dd$1 | 00$d$0$dd$1 | 22 | 17 |
|   |   | 00$d$+$dd$0 | +−$d$−$dd$0 | +−$d$−$dd$0 | −+$d$+$dd$0 | −+$d$+$dd$0 | −+$d$+$dd$0 | −+$d$+$dd$− | +−$d$−$dd$+ |

models are linearly separable and include (X4, X6). We select 11 models more without X6 although eight models are dropped from the table. Therefore, these models are not linearly separable. The "model" column shows the suffix of variables. Each model, except for LDF and Logistic, has two expressions by a given symbol. The lower row is the original coefficient, and the upper row is the modified coefficient by setting the intercept to one. The rule of a given symbols is as follows:

1. Symbols "+," "−," and "0" show that the coefficients are positive (lower limit of the 95 % CI > 0), negative (upper limit of the 95 % CI < 0), and zero (the 95 % CI includes 0) at the 5 % significant level, respectively. If the model has the symbol "0," we should not choose this model by the coefficient standard.
2. Symbol "$d$" means that the variable has been dropped from the model.
3. Symbol "$Z$" means that 100 coefficients are zeroes that means natural feature-selection for 100 training samples. If the model has the symbol "$Z$," we consider this model to be redundant. Moreover, this model is the same as the model that dropped variables with symbol "$d$."
4) Symbol "1" means that 100 intercepts are ones. Symbol "*" means that the intercept is 1/0 because the same original intercepts are zero.

### 6.3.2.2 Revised IP-OLDF

Equation (6.1) is the full model for Revised IP-OLDF. We represent the full model as the symbol "+ZZ++−Z" in Table 6.6. The symbols for X2 and X3, and the intercept are "Z." This fact indicates that we can drop these two variables from the full model and that the full model is redundant. If the symbol for the intercept is "Z" in the second row, the first row is the same as the second row because we need not divide by the original intercept. Equation (6.2) shows the third five-variable model (X1, X3, X4, X5, X6). We select this model as the best model for the minimum $M2$ standard. The symbol is "+dZ++−Z." Equation (6.3) shows the fourth five-variable model (X1, X2, X4, X5, X6). The symbol is "+Zd++−Z." Equation (6.4) shows the eighth four-variable model (X1, X4, X5, X6). The symbol is "+dd++−Z." Because we believe that the symbols "Z" and "$d$" initially have the same effect, these four models are equivalent. However, the four $M2$s of the four models are different, such as 0.30, 0.26, 0.30, and 0.27 %, because the 95 % CI of each coefficient is different. We can determine that X2 and X3 are less significant among the six variables.

$$\begin{aligned} SN = 1 : RIP &= 1.037 \times X1 + Z \times X2 + Z \times X3 + 2.197 \times X4 + 2.285 \times X5 - 1.812 \times X6 + Z \\ &= 1.037 \times X1 + 2.197 \times X4 + 2.285 \times X5 \quad - 1.812 \times X6. \\ &\quad [0.147, 1.455] \ [0.878, 3.729] \ [0.539, 4.278] \ [-2.438, -0.556] \end{aligned}$$

(6.1)

## 6.3 100-Fold Cross-Validation for Small Sample Method

$$\begin{aligned}
\text{SN} = 3 : \text{RIP} &= 1.037 \times X1 + d \times X2 + Z \times X3 + 2.197 \times X4 + 2.292 \times X5 - 1.908 \times X6 + Z \\
&= 1.037 \times X1 + 2.197 \times X4 + 2.292 \times X5 - 1.908 \times X6 \\
&\quad [0.231, 1.353]\ [0.878, 3.124]\ [0.539, 4.049]\ [-2.317, -0.612]
\end{aligned} \tag{6.2}$$

$$\begin{aligned}
\text{SN} = 4 : \text{RIP} &= 1.037 \times X1 + Z \times X2 + d \times X3 + 2.20 \times X4 + 2.30 \times X5 - 1.84 \times X6 + Z \\
&= 1.037 \times X1 + 2.200 \times X4 + 2.300 \times X5 \quad - 1.840 \times X6 \\
&\quad [0.147, 1.455]\ [0.878, 3.729]\ [0.539, 4.278]\ [-2.438, -0.556]
\end{aligned} \tag{6.3}$$

$$\begin{aligned}
\text{SN} = 8 : \text{RIP} &= 1.037 \times X1 + d \times X2 + d \times X3 + 2.197 \times X4 + 2.3 \times X5 - 1.84 \times X6 + Z \\
&= 1.037 \times X1 + 2.197 \times X4 \quad + 2.3 \times X5 \quad - 1.84 \times X6 \\
&\quad [0.231, 1.297]\ [0.878, 3.192]\ [0.659, 4.242]\ [-2.314, -0.612]
\end{aligned} \tag{6.4}$$

Equation (6.5) is the original 12th three-variable model, with symbol "*ddd++−0.*" Shinmura (2015a) showed the excellent result of setting the intercept to one by dividing the original coefficients. We divide each original coefficient by (the intercept (143.36) + 0.000001); it then becomes "*ddd++−\**" in Eq. (6.6). Because fewer than 25 intercepts are zero, the intercepts have 1/0 values. We denote this status as the symbol "\*." We know the values of coefficients are quite different in Eqs. (6.1), (6.2), (6.3), and (6.4). Equation (6.7) is the 16th two-variable model, with symbol "*ddd+d-+*". If we divide the coefficients by the original intercept, we obtain Eq. (6.8), denoted by "*ddd+d-*1." The symbol "1" means that all intercepts are one. To this point, we have not been able to understand the useful meaning of the 95 % CI of discriminant coefficients. In this research, we investigate 16 linearly separable models and find that all the coefficients of six models have positive or negative values, but not zero. Therefore, $(4, 6) \subset (4 - 6) \subset (1, 4 - 6) \subset (1, 2, 4 - 6)/(1, 3, 4 - 6) \subset (1 - 6)$, are more valuable than other ten models. The result of the "minimum $M2$ standard" and "coefficient standard" matches well only for six linearly separable models of Revised IP-OLDF. We claim that the "coefficient standard" supports the best model.

$$\begin{aligned}
\text{SN} = 12 : \text{RIP} &= 4.346 \times X4 + 5.432 \times X5 - 1.498 \times X6 + 143.356 \\
&\quad [0.80, 7.71]\ [0.88, 11.87]\ [-2.60, -0.44]\ [0, 251]
\end{aligned} \tag{6.5}$$

$$\begin{aligned}
&= 0.023 \times X4 + 0.033 \times X5 - 0.012 \times X6 + 1/0 \\
&\quad [0.001, 7E6]\ [0.01, 1E7]\ [-1E6, -0.01] \quad *
\end{aligned} \tag{6.6}$$

$$\begin{aligned}
\text{SN} = 16 : \text{RIP} &= 3.846 \times X4 - 5.321 \times X6 + 699.395 \\
&\quad [0.364, 44]\ [-48, -2.54]\ [345, 6348]
\end{aligned} \tag{6.7}$$

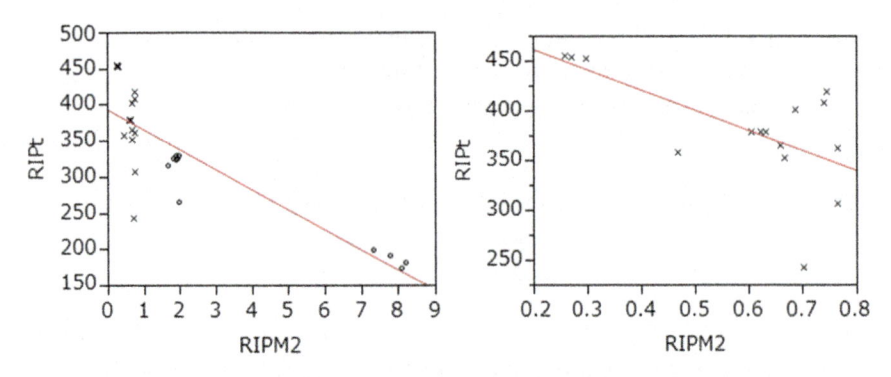

**Fig. 6.2** Scatter plots for M2 versus t-value (*left r* = −0.8449, *right r* = −0.6678)

$$\begin{matrix} =0.007 \times X4 & -0.008 \times X6 + 1 \\ [0.001, 0.01] & [-0.008, -0.007] \end{matrix} \quad (6.8)$$

We calculate the discriminant scores by Revised IP-OLDF and calculate the *t*-value of the two classes listed in Table 6.5. The *t*-value of the best model is 456, and it is the maximum value among 27 discriminant scores in Table 6.2 and 6.3. Figure 6.2 shows the scatter plots. The x-axis is *M*2s (RIPM2), and the y-axis is *t*-values (RIPt). The left plot has 27 points, and their correlation is $r = -0.84$. The symbol "×" is assigned to 16 linearly separable models and symbol "." to 11 models. The right plot has only 16 linearly separable models, and the correlation is $r = -0.67$. Although the best model (1, 3–6) of RIP has the maximum *t*-value (Maximum *t*-value Standard), we are concerned that the *t* test always supports the best model of RIP.

We can confirm that the coefficients of all 11 models include "0." Although we can observe the second best model among these 11 models in Table 6.6, we need not consider these models by the coefficient standard.

### 6.3.2.3 Hard-Margin SVM (H-SVM) and Other LDFs

Equation (6.9) is the full model of H-SVM, and its symbol is "000++−0." In order to avoid division by zero, we divide the coefficients by the original intercept ((−102.38) + 0.000001). Following this notation rule, we can represent Eq. (6.10) as the symbol "0000001." In Table 6.6, this equation appears in the first row, and its original symbol is in the second row because we believe that the model with intercept equal to one is better than the initial coefficient. The signs of the 12th and 16th models are "*ddd*++−1" and "*ddd*+*d*−1," respectively. Although the coefficient standard supports the 12th and 16th models of H-SVM, the maximum *t*-value standard support the sixth model. We predict that H-SVM is superior to Revised IP-OLDF for LSD because the support vector (SV) is efficient for LSD. On the other hand, we predict that Revised IP-OLDF might overestimate LSD because the

6.3 100-Fold Cross-Validation for Small Sample Method

**Fig. 6.3** Scatter plots for *M2* versus *t*-value ($r = -0.6871$)

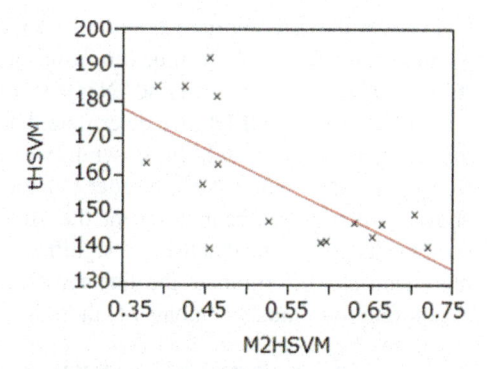

former might arbitrarily search for one of the interior points of OCP. From our research, we assume that Revised IP-OLDF searches for the OCP gravity that causes the stability results explained in Chap. 7. Figure 6.3 shows the scatter plot of 16 models. The correlation is $r = -0.69$.

$$\begin{aligned}
\text{H-SVM} = &0.993 \times X1 \quad +0.567 \times X2 \quad +0.276 \times X3 + 1.645 \times X4 \\
&[-0.57, 1.74] \quad [-0.54, 1.41] \quad [-1.06, 0.83] \quad [0.37, 2.54] \\
&+1.391 \times X5 \quad -1.635 \times X6 \quad -102.38 \\
&[0.49, 3.02] \quad [-2.17, -0.59] \quad [-456, 285]
\end{aligned} \tag{6.9}$$

$$= -0.004 \times X1 - 0.003 \times X2 + 0.0002 \times X3 - 0.004 \times X4 - 0.004 \times X5 + 0.01*X6 + 1$$
$$[-0.03, 0.13]\ [-0.02, 0.05]\ [-0.03, 0.07]\ \ [-0.1, 0.3]\ [-0.11, 0.24]\ [-0.27, 0.09]$$
$$\tag{6.10}$$

The coefficients of the 12th, 16th, and 22nd models of SVM4 and SVM1 do not include the symbol "0." We need not discuss these models because the values of *M2*s are larger than those of Revised IP-OLDF. Although Revised LP-OLDF and Revised IPLP-OLDF select the 16th model, we need not discuss these models for the same reason. Because Fisher's LDF and logistic regression by JMP script (Sall et al. 2004) do not output 100 discriminant coefficients, we discriminate the original data by the regression analysis and logistic regression. In Table 6.6, the first rows of Fisher's LDF and logistic regression show NMs by the original Swiss banknote data. If we divide this number by two, the calculated value is the error rate because the sample size is 200 cases. The full models of Fisher's LDF and logistic regression are Eqs. (6.11) and (6.12). The numbers enclosed in parentheses are the standard errors although square bracket ("[]") means 95 % CI in Eq. (6.1). Coefficients right shoulder * is the variable rejected at the 5 % level. In Eq. (6.11) and (6.12), the multiplication sign "×" is omitted in order to avoid the mistake of the symbols "*" and "×." The symbols for Fisher's LDF and logistic regression are "0+- - -+-" and "0000000," respectively. Table 6.6 lists the 95 % CI that supports the eight models of Fisher's LDF. However, Fisher's LDF never discriminates the 16 linearly separable models correctly. Logistic regression calculates SE from the Hessian matrix.

Such SE values are enormous, and all CI include zero. Therefore, JMP outputs a warning message, such "estimation is unstable" for the linearly separable model (Firth 1993). However, if we find "NM = 0" on the ROC by JMP output and "MNM = 0" by Revised IP-OLDF, we determine that logistic regression can discriminate a linearly separable model. In general, we recommend using the exact logistic regression supported by SAS in order to avoid complex work such as the one given above. Although we cannot accept the 16 linearly separable models because all coefficients are accepted at the 5 % significant level and are zeroes, we judge logistic regression can discriminate the linearly separable models correctly. Although we accept only two models among 11, there is no meaning for the discrimination.

$$\text{LDF} = -0.03 X1 + 0.23^* X2 - 0.22^* X3 - 0.30^* X4 - 0.31^* X5 + 0.42^* X6 - 47.18^*$$
$$\quad (0.06) \quad (0.09) \quad (0.08) \quad (0.02) \quad (12.18) \quad (0.03) \quad (13.10)$$
$$(6.11)$$

$$\text{Logistic} = 30.33 X1 - 3.36 X2 + 4.86 X3 + 36.69 X4 + 50.72 X5 - 28.63 X6 - 3594.33.$$
$$\quad (28411) \quad (35162) \quad (48244) \quad (8772) \quad (18142) \quad (8954) \quad (8608558)$$
$$(6.12)$$

## 6.4 Explanation 1 for Swiss Banknote Data

### 6.4.1 Matroska in Linearly Separable Data

When we discriminated Swiss banknote data by IP-OLDF (Shinmura 1998, 2000a, b), we found that MNM of the two-variable model (X4, X6) is zero. Although there are 63 ($=2^6 - 1$) models in Table 6.7, 16 models, including (X4, X6), are zero as listed in Table 6.5, and the other MNMs of 47 models are not zero because these models do not include (X4, X6). We did not understand that LSD has the Matroska structure, which is a linearly separable model. The largest Matroska includes smaller Matroska in it. Because there are 16 Matroskas, we can produce Matroska products by a combination of the smaller 15 Matroskas into the largest Matroska. It is the most important fact that the two-variable model (X4, X6) can completely explain the structure of 16 models. We call the two-variable model (X4, X6) "the basic gene space (BGS)" in the gene analysis.

Table 6.8 lists the structure of the Swiss banknote data from the perspective of the Matroska producer. Column "SN2" is the product number of Matroska products. Five columns, such "6," "5," "4," "3," and "2," are the Matroska size. The largest Matroska includes four five-variable linearly separable models (smaller Matroska), which are (X2, X3, X4, X5, X6), (X1, X3, X4, X5, X6), (X1, X2, X4, X5, X6), and (X1, X2, X3, X4, X6). Because two models, such as (X1–X3, X5, X6) and (X1–X5), are not linearly separable, these models are not Matroska. Each five-variable Matroska includes three four-variable Matroska, and each

## 6.4 Explanation 1 for Swiss Banknote Data

**Table 6.7** MNM and NMs of 16 linearly separable models

| SN | $p$ | Var. | RIP | SVM1 | LDF | QDF | RDA |
|---|---|---|---|---|---|---|---|
| 1 | 6 | 1–6 | 0 | 1 | 1 | 1 | 1 |
| 2 | 5 | 2–6 | 0 | 1 | 1 | 1 | 1 |
| 3 | 5 | 1, 3–6 | 0 | 1 | 1 | 1 | 1 |
| 4 | 5 | 1, 2, 4–6 | 0 | 1 | 1 | 1 | 1 |
| 5 | 5 | 1–4, 6 | 0 | 2 | 1 | 1 | 1 |
| 6 | 4 | 3–6 | 0 | 1 | 1 | 1 | 1 |
| 7 | 4 | 2, 4–6 | 0 | 1 | 1 | 1 | 1 |
| 8 | 4 | 1, 4–6 | 0 | 1 | 1 | 1 | 1 |
| 9 | 4 | 2–4, 6 | 0 | 2 | 1 | 1 | 1 |
| 10 | 4 | 1, 3, 4, 6 | 0 | 2 | 1 | 1 | 1 |
| 11 | 4 | 1, 2, 4, 6 | 0 | 2 | 2 | 1 | 1 |
| 12 | 3 | 4–6 | 0 | 1 | 1 | 1 | 1 |
| 13 | 3 | 3, 4, 6 | 0 | 2 | 1 | 1 | 1 |
| 14 | 3 | 1, 4, 6 | 0 | 2 | 2 | 2 | 1 |
| 15 | 3 | 2, 4, 6 | 0 | 2 | 1 | 1 | 1 |
| 16 | 2 | 4, 6 | 0 | 2 | 3 | 1 | 1 |

**Table 6.8** MNM of 16 linearly separable models

| SN2 | 6 | 5 | 4 | 3 | 2 |
|---|---|---|---|---|---|
| 1 | 1–6 | 2–6 | 3–6 | 4–6 | 4, 6 |
| 2 | | | | 3, 4, 6 | 4, 6 |
| 3 | | | 2, 4–6 | 4–6 | 4, 6 |
| 4 | | | | 2, 4, 6 | 4, 6 |
| 5 | | | 2–4, 6 | 3, 4, 6 | 4, 6 |
| 6 | | | | 2, 4, 6 | 4, 6 |
| 7 | | 1, 3–6 | 3–6 | 4–6 | 4, 6 |
| 8 | | | | 3, 4, 6 | 4, 6 |
| 9 | | | 1, 4–6 | 4–6 | 4, 6 |
| 10 | | | | 1, 4, 6 | 4, 6 |
| 11 | | | 1, 3, 4, 6 | 3, 4, 6 | 4, 6 |
| 12 | | | | 1, 4, 6 | 4, 6 |
| 13 | | 1, 2, 4–6 | 2, 4–6 | 4–6 | 4, 6 |
| 14 | | | | 2, 4, 6 | 4, 6 |
| 15 | | | 1, 4–6 | 4–6 | 4, 6 |
| 16 | | | | 1, 4, 6 | 4, 6 |
| 17 | | | 1, 2, 4, 6 | 2, 4, 6 | 4, 6 |
| 18 | | | | 1, 4, 6 | 4, 6 |
| 19 | | 1–4, 6 | 2–4, 6 | 3, 4, 6 | 4, 6 |
| 20 | | | | 2, 4, 6 | 4, 6 |
| 21 | | | 1, 3, 4, 6 | 3, 4, 6 | 4, 6 |
| 22 | | | | 1, 4, 6 | 4, 6 |
| 23 | | | 1, 2, 4, 6 | 2, 4, 6 | 4, 6 |
| 24 | | | | 1, 4, 6 | 4, 6 |

four-variable Matroska consists of two three-variable Matroska. At last, each three-variable Matroska includes one two-variable Matroska, such as the same ($X4$, $X6$). We call the smallest Matroska, such as ($X4$, $X6$), "BGS" using the terminology of microarray dataset in Chap. 8. By Method 2, we can conclude that the structure of Swiss banknote data is as follows:

1. LSD has the Matroska structure, and we call all linearly separable models, "Matroska." The full model of the data with six variables is the largest Matroska that includes smaller Matroska from five to two variables.
2. The two-variable Matroska, such as ($X4$, $X6$), is unique BGS in the terminology of Method 2. We can understand the structure of LSD by BGS completely because all Matroska must include BGS by the monotonic decrease of MNM.
3. The Matroska producer can make 24 Matroska products by the combination of 15 Matroska.
4. We call each Matroska product, "the Matroska series." For example, the first "Matroska product SN = 1" has the following Matroska series: (1–6) $\ni(2-6)\ni(3-6)\ni(4-6)\ni(4,6)$. We can produce the Matroska product with the instructions of the Matroska series.

### 6.4.2 Explanation 1 of Method 2 by Swiss Banknote Data

We explain the Method 2 with the Swiss banknote data. Table 6.9 lists the coefficients of Revised IP-OLDF. Because the first 16 models include ($X4$, $X6$), these models are MNM = 0. The seven columns in Table 6.9 (from "$X1$" to "$c$") are the coefficients of Revised IP-OLDF. When we discriminate the data by Revised IP-OLDF, the two coefficients of $X2$ and $X3$ become zero naturally. Therefore, we can make feature-selection naturally from six variables to four variables. Next, when we discriminate "four-variable model (1, 4–6) in row SN = 8," we cannot reduce the four-variable model to a smaller model. Therefore, we stop the Method 2 and call this four-variable model, "the small Matroska (SM)." Because MP solver finds the global solution, it outputs the result by "first-in, first-out" rule, and it cannot find BGS that is one of global solution having the same value of MNM. After this step, we must survey BGS, such as ($X4$, $X6$), with the statistical approach (Shinmura 2016a, b). After deleting BGS from the full model, we discriminate the four-variable model ($X1$–$X3$, $X5$) by Revised IP-OLDF. Because MNM of this model is 18, we stop Method 2 and conclude that we have found one BGS, such as ($X4$, $X6$) in the Swiss banknote data, through a manual survey. In the near future, we would like to develop the Revised LINGO Program 3 of Method 2 to find all BGS automatically.

Table 6.10 lists the coefficients of H-SVM. All the coefficients of the full models are not zero. Therefore, we cannot naturally make feature-selection by H-SVM. We also cannot make feature-selection naturally by SVM4. Although many researchers erroneously believe that S-SVM can discriminate the linearly separable model

6.4 Explanation 1 for Swiss Banknote Data

**Table 6.9** Six NMs and coefficients of Revised IP-OLDF

| SN | MNM | Var. | X1 | X2 | X3 | X4 | X5 | X6 | c |
|---|---|---|---|---|---|---|---|---|---|
| 1 | 0 | 1–6 | −1 | 0 | 0 | −3 | −3 | 2 | 0 |
| 2 | 0 | 2–6 | | 0 | 2 | −4 | −5 | 2 | −513 |
| 3 | 0 | 1, 3–6 | −1 | | 0 | −3 | −3 | 2 | 0 |
| 4 | 0 | 1, 2, 4–6 | −1 | 0 | | −3 | −3 | 2 | 0 |
| 5 | 0 | 1–4, 6 | 7 | −5 | 2 | −11 | | 11 | −2606 |
| 6 | 0 | 3–6 | | | 2 | −4 | −4 | 2 | −444.3 |
| 7 | 0 | 2, 4–6 | | −3 | | −2 | −3 | 2 | 113.14 |
| 8 | 0 | 1, 4–6 | −1 | | | −3 | −3 | 2 | 0 |
| 9 | 0 | 2–4, 6 | | −5 | 7 | −22 | | 24 | −3408 |
| 10 | 0 | 1, 3, 4, 6 | 14 | | −10 | −28 | | 28 | −5308 |
| 11 | 0 | 1, 2, 4, 6 | 8 | −4 | | −14 | | 14 | −3126 |
| 12 | 0 | 4–6 | | | | −5 | −6 | 3 | −250.7 |
| 13 | 0 | 3, 4, 6 | | | 0 | −44 | | 48 | −6348 |
| 14 | 0 | 1, 4, 6 | 0 | | | −44 | | 48 | −6348 |
| 15 | 0 | 2, 4, 6 | | 0 | | −44 | | 48 | −6348 |
| 16 | 0 | 4, 6 | | | | −44 | | 48 | −6348 |
| (1–6) - BGS | 18 | 1–3, 5 | 9506.8 | −4625 | −9990 | 0 | −7071 | 0 | −67712 |

**Table 6.10** Coefficients of H-SVM

| SN | Var. | X1 | X2 | X3 | X4 | X5 | X6 | c |
|---|---|---|---|---|---|---|---|---|
| 1 | 1–6 | −1.14 | −0.57 | 0.124 | −2.3 | −2.8 | 1.796 | 102 |
| 2 | 2–6 | | −2.08 | 0.746 | −2.62 | −2.49 | 2.275 | −94 |
| 3 | 1, 3–6 | −1.48 | | −0.1 | −2.27 | −2.89 | 1.699 | 146 |
| 4 | 1, 2, 4–6 | −1.3 | −0.29 | | −2.3 | −2.83 | 1.758 | 123 |
| 5 | 1–4, 6 | 7.694 | −5.54 | 3.746 | −10.3 | | 10.19 | −2759 |
| 6 | 3–6 | | | 0.657 | −3.31 | −2.86 | 3.018 | −447 |
| 7 | 2, 4–6 | | −1.92 | | −3.17 | −2.12 | 2.452 | −42 |
| 8 | 1, 4–6 | −1.45 | | | −2.27 | −2.9 | 1.738 | 120 |
| 9 | 2–4, 6 | | −4.83 | 6.897 | −21.5 | | 23.72 | −3408 |
| 10 | 1, 3, 4, 6 | 13.66 | | −9.76 | −28 | | 27.7 | −5308 |
| 11 | 1, 2, 4, 6 | 8.464 | −4.23 | | −14.4 | | 14.14 | −3126 |
| 12 | 4–6 | | | | −3.75 | −2.5 | 3.125 | −377 |
| 13 | 3, 4, 6 | | | 0 | −44 | | 48 | −6348 |
| 14 | 1, 4, 6 | 24.75 | | | −29.2 | | 30.68 | −9366 |
| 15 | 2, 4, 6 | | 0 | | −44 | | 48 | −6348 |
| 16 | 4, 6 | | | | −44 | | 48 | −6348 |

exactly and prefer to select a small value, such as $c = 1$, as penalty $c$, only SVM1 (for $c = 1$) cannot discriminate the 16 modes as linearly separable. We confirm that SVM4 is better than SVM1 in many trials. However, only H-SVM and Revised IP-OLDF can discriminate the linearly separable model theoretically. Therefore, we can omit to use other LDFs.

Table 6.11 lists the coefficients of Revised LP-OLDF. The two coefficients of $X2$ and $X3$ are zero. Therefore, when we discriminate the four-variable model ($X1$, $X4$, $X5$, $X6$), none of the coefficients are zero, and we stop Method 2 at the four-variable model. Although Revised LP-OLDF and Revised IPLP-OLDF can make feature-selection naturally, we never use these OLDFs for the Theory 2 because they cannot recognize LSD theoretically. Moreover, these OLDFs cannot reduce the high-dimensional microarray dataset to the small Matroska drastically, as explained in Chap. 8. I cannot explain this reason. It is one of future works. Howevere, it may be caused by the branch and bound algorithm of LINGO IP-solver.

**Table 6.11** Coefficients of Revised LP-OLDF

| SN | Var. | X1 | X2 | X3 | X4 | X5 | X6 | c |
|---|---|---|---|---|---|---|---|---|
| 1 | 1–6 | −1.09 | 0 | 0 | −2.61 | −2.83 | 2.06 | 0 |
| 2 | 2–6 |  | −2.94 | 0 | −2.47 | −2.7 | 2.3 | 113.135 |
| 3 | 1, 3–6 | −1.09 |  | 0 | −2.61 | −2.83 | 2.06 | 0 |
| 4 | 1, 2, 4–6 | −1.09 | 0 |  | −2.61 | −2.83 | 2.06 | 0 |
| 5 | 1–4, 6 | 7.222 | −5.24 | 2.331 | −11.1 |  | 10.91 | −2605.6 |
| 6 | 3–6 |  |  | 0 | −4.8 | −6.48 | 2.6 | −250.69 |
| 7 | 2, 4–6 |  | −2.94 |  | −2.47 | −2.7 | 2.3 | 113.135 |
| 8 | 1, 4–6 | −1.09 |  |  | −2.61 | −2.83 | 2.06 | 0 |
| 9 | 2–4, 6 |  | −4.83 | 6.897 | −21.5 |  | 23.72 | −3408.1 |
| 10 | 1, 3, 4, 6 | 13.66 |  | −9.76 | −28 |  | 27.7 | −5307.6 |
| 11 | 1, 2, 4, 6 | 8.464 | −4.23 |  | −14.4 |  | 14.14 | −3126.5 |
| 12 | 4–6 |  |  |  | −4.8 | −6.48 | 2.6 | −250.69 |
| 13 | 3, 4, 6 |  |  | 0 | −44 |  | 48 | −6347.8 |
| 14 | 1, 4, 6 | 0 |  |  | −44 |  | 48 | −6347.8 |
| 15 | 2, 4, 6 |  | 0 |  | −44 |  | 48 | −6347.8 |
| 16 | 4, 6 |  |  |  | −44 |  | 48 | −6347.8 |
| (1–6) - BGS | 1–3, 5 | 1.914 | −1.22 | −2.7 | 0 | −1.53 | 0 | 115.031 |

## 6.5 Summary

Table 6.12 lists the comparison of the six models selected by Revised IP-OLDF. Because SVM1 and Fisher's LDF cannot discriminate the linearly separable models theoretically, there is no need to compare among the eight LDFs. We omit Revised IPLP-OLDF and Revised LP-OLDF because these OLDFs are inferior to Revised IP-OLDF. Column "M2" shows the value of $M2$. The number before the colon indicates the rank of the useful models. The third model of Revised IP-OLDF has the minimum value, 0.26, among all models by eight LDFs and the maximum value, 456, of the t test. If we compare the third model of Revised IP-OLDF with H-SVM, SVM4, and logistic regression, we can confirm that the third model of Revised IP-OLDF is the best model. Both the coefficient and $t$ test standards support the "minimum $M2$ standard." To this point, we cannot determine the best model uniquely. Even if Fisher had developed SE for the error rate and discriminant coefficients, we could not choose the best model uniquely. Because we can now obtain powerful computer power and user-friendly solvers, such as LINGO (Schrage 2006) and JMP, we should develop a new theory of discriminant analysis through a computer-intensive approach. Most statisticians respect Fisher. He opened a new frontier for much of the statistical theory through his intellectual consideration without computer power. Therefore, we believe that no researcher might discuss our claim seriously. However, we can use powerful computer power and software, such as statistical software JMP and MP solver LINGO. We are fortunate to be a generation with advanced technology, unlike Fisher era. We should develop analytical techniques that are tailored to the characteristics of individual data without normal distribution.

**Table 6.12** Comparison of six models

| SN | Model | RIP | | | H-SVM | | | SVM4 | | Logistic |
|---|---|---|---|---|---|---|---|---|---|---|
| | | M2 | coeff. | t | M2 | coeff. | t | M2 | coeff. | M2 |
| 1 | 1–6 | 4: 0.30 | +ZZ++ −Z | 4: 453 | 6: 0.53 | 0000001 | 5: −147 | 6: 0.52 | 0000001 | 6: 0.55 |
| 3 | 1, 3–6 | 1: **0.26** | +dZ++ −Z | 1: **456** | 5: 0.46 | 0d00001 | 3: −163 | 4: 0.46 | 0d00001 | 4: 0.52 |
| 4 | 1, 2, 4–6 | 3: 0.30 | +Zd++ −Z | 3: 453 | 3: 0.45 | 00d0001 | 4: −158 | 3: 0.45 | 00d0001 | 5: 0.55 |
| 8 | 1, 4–6 | 2: 0.27 | +dd++ −Z | 2: 454 | 1: 0.38 | 0dd0001 | 2: −164 | 1: 0.37 | 0dd0001 | 2: 0.46 |
| 12 | 4–6 | 6: 0.69 | ddd++ −* | 5: 402 | 2: 0.39 | ddd++ −1 | 1: 184 | 2: 0.39 | ddd++ −1 | 1: 0.41 |
| 16 | 4, 6 | 5: 0.47 | ddd+d −1 | 6: 359 | 4: 0.46 | ddd+d −1 | 6: 140 | 5 0.46 | ddd+d −1 | 3: 0.47 |

# References

Buhlmann P, Geer AB (2011) Statistics for high-dimensional data-method, theory and applications. Springer, Berlin

Cox DR (1958) The regression analysis of binary sequences (with discussion). J Roy Stat Soc B 20:215–242

Firth D (1993) Bias reduction of maximum likelihood estimates. Biometrika 80:27–39

Fisher RA (1936) The use of multiple measurements in taxonomic problems. Ann Eugenics 7:179–188

Fisher RA (1956) Statistical methods and statistical inference. Hafner Publishing Co, New Zealand

Flury B, Rieduyl H (1988) Multivariate statistics: a practical approach. Cambridge University Press, Cambridge

Friedman JH (1989) Regularized discriminant analysis. J Am Stat Assoc 84(405):165–175

Goodnight JH (1978) SAS technical report—the sweep operator: its importance in statistical computing—(R100). SAS Institute Inc, USA

Jeffery IB, Higgins DG, Culhane C (2006) Comparison and evaluation of methods for generating differentially expressed gene lists from microarray data. BMC Bioinf 7:359:1–16. doi: 10.1186/1471-2105-7-359

Lachenbruch PA, Mickey MR (1968) Estimation of error rates in discriminant analysis. Technometrics 10:1–11

Miyake A, Shinmura S (1976) Error rate of linear discriminant function. In: Gremy F (ed) Dombal FT. North-Holland Publishing Company, pp 435–445

Miyake A, Shinmura S (1979) An algorithm for the optimal linear discriminant functions. Proceedings of the international conference on cybernetics and society, pp 1447–1450

Sall JP (1981) SAS regression applications. SAS Institute Inc., USA. (Shinmura S. translate Japanese version)

Sall JP, Creighton L, Lehman A (2004) JMP start statistics, third edition. SAS Institute Inc., USA. (Shinmura S. edits Japanese version)

Schrage L (1991) LINDO—an optimization modeling systems. The Scientific Press, UK. (Shinmura S. & Takamori, H. translate Japanese version)

Schrage L (2006) Optimization modeling with LINGO. LINDO Systems Inc., USA. (Shinmura S. translates Japanese version)

Shinmura S (1998) Optimal linear discriminant functions using mathematical programming. J Jpn Soc Comput Stat 11/2:89–101

Shinmura S, Tarumi T (2000) Evaluation of the optimal linear discriminant functions using integer programming (IP-OLDF) for the normal random data. J Jpn Soc Comput Stat 12(2):107–123

Shinmura S (2000a) A new algorithm of the linear discriminant function using integer programming. New Trends Prob Stat 5:133–142

Shinmura S (2000b) Optimal linear discriminant function using mathematical programming. Dissertation, March 200:1–101, Okayama University, Japan

Shinmura S (2003) Enhanced algorithm of IP-OLDF. ISI2003 CD-ROM, pp 428–429

Shinmura S (2004) New algorithm of discriminant analysis using integer programming. IPSI 2004 Pescara VIP Conference CD-ROM, pp 1–18

Shinmura S (2005) New age of discriminant analysis by IP-OLDF –beyond Fisher's linear discriminant function. ISI2005, pp 1–2

Shinmura S (2007) Overviews of discriminant function by mathematical programming. J Jpn Soc Comput Stat 20(1-2):59–94

Shinmura S (2010a) The optimal linearly discriminant function (Saiteki Senkei Hanbetu Kansuu). Union of Japanese Scientist and Engineer Publishing, Japan

Shinmura S (2010b) Improvement of CPU time of Revised IP-OLDF using Linear Programming. J Jpn Soc Comput Stat 22(1):39–57

# References

Shinmura S (2011a) Beyond Fisher's linear discriminant analyisi—new world of the discriminant analysis. ISI CD-ROM, pp 1–6

Shinmura S (2011b) Problems of discriminant analysis by mark sense test data. Jpn Soc Appl Stat 40(3):157–172

Shinmura S (2013) Evaluation of optimal linear discriminant function by 100-fold cross-validation. ISI CD-ROM, pp 1–6

Shinmura S (2014a) End of discriminant functions based on variance-covariance matrices. ICORE2014, pp 5–16

Shinmura S (2014b) Improvement of CPU time of linear discriminant functions based on MNM criterion by IP. Stat Optim Inf Comput 2:114–129

Shinmura S (2014c) Comparison of linear discriminant functions by $k$-fold cross-validation. Data Anal 2014:1–6

Shinmura S (2015a) The 95 % confidence intervals of error rates and discriminant coefficients. Stat Optim Inf Comput 2:66–78

Shinmura S (2015b) A trivial linear discriminant function. Stat Optim Inf Comput 3:322–335. doi:10.19139/soic.20151202

Shinmura S (2015c) Four serious problems and new facts of the discriminant analysis. In: Pinson E, Valente F, Vitoriano B (ed) Operations research and enterprise systems, pp 15–30. Springer, Berlin (ISSN: 1865-0929, ISBN: 978-3-319-17508-9, doi:10.1007/978-3-319-17509-6)

Shinmura S (2015d) Four problems of the discriminant analysis. ISI 2015:1–6

Shinmura S (2015e) The discrimination of microarray data (Ver. 1). Res Gate 1:1–4. 28 Oct 2015

Shinmura S (2015f) Feature selection of three microarray data. Res Gate 2:1–7. 1 Nov 2015

Shinmura S (2015g) Feature Selection of Microarray Data (3)—Shipp et al. Microarray Data. Research Gate (3), 2015: 1–11

Shinmura S (2015h) Validation of feature selection (4)—Alon et al. microarray data. Res Gate (4), 2015, pp 1–11

Shinmura S (2015i) Repeated feature selection method for microarray data (5). Res Gate 5:1–12. 9 Nov 2015

Shinmura S (2015j) Comparison Fisher's LDF by JMP and revised IP-OLDF by LINGO for microarray data (6). Res Gate 6:1–10. 11 Nov 2015

Shinmura S (2015k) Matroska trap of feature selection method (7)—Golub et al. microarray data. Res Gate (7), 18, 2015, pp 1–14

Shinmura S (2015l) Minimum Sets of Genes of Golub et al. Microarray Data (8). Research Gate (8) 1–12. 22 Nov 2015

Shinmura S (2015m) Complete lists of small matroska in Shipp et al. microarray data (9). Res Gate (9), pp 1–81

Shinmura S (2015n) Sixty-nine small matroska in Golub et al. microarray data (10). Res Gate, pp 1–58

Shinmura S (2015o) Simple structure of Alon et al. microarray data (11). Res Gate (1.1), pp 1–34

Shinmura S (2015p) Feature selection of Singh et al. microarray data (12). Res Gate (12), pp 1–89

Shinmura S (2015q) Final list of small matroska in Tian et al. microarray data. Res Gate (13), pp 1–160

Shinmura S (2015r) Final list of small matroska in Chiaretti et al. microarray data. Res Gate (14), pp 1–16

Shinmura S (2015s) Matroska feature selection method for microarray data. Res Gate (15), pp 1–16

Shinmura S (2016a) The best model of swiss banknote data. Stat Optim Inf Comput, 4:118–131. International Academic Press (ISSN: 2310-5070 (online) ISSN: 2311-004X (print), doi: 10.19139/soic.v4i2.178)

Shinmura S (2016b) Matroska feature-selection method for microarray data. Biotechnology 2016:1–8

Shinmura S (2016c) discriminant analysis of the linear separable data—Japanese-automobiles. J Stat Sci Appl X X:0–14

Simon N, Friedman J, Hastie T, Tibshirani R (2013) A sparse-group lasso. J Comput Graph Stat 22:231–245

Vapnik V (1995) The nature of statistical learning theory. Springer, Berlin

# Chapter 7
# Japanese-Automobile Data
## Explanation 2 of Matroska Feature-Selection Method (Method 2)

## 7.1 Introduction

One of our master's course students presented the statistical report about "The Japanese-automobile Data (Shinmura 2015c, 2016c)" instead of the test. There are 29 regular and 15 small cars with six independent variables, such as the emission rate ($X1$), price ($X2$), number of seats ($X3$), $CO_2$ ($X4$), fuel ($X4$), and sales ($X6$). The student's main theme was the prediction of $X6$ by the other five variables. However, the result was not clear. First, we used these data for the research of data envelopment analysis (DEA).[1] Next, we found three points that are critical for discriminant analysis:

1. Discrimination of linearly separable data (LSD)

We can easily recognize that these data are LSD because $X1$ and $X3$ can separate two classes completely by two box–whisker plots. They are two basic gene sets (BGSs) in Chap. 8.

2. Problem 3

The forward stepwise procedure selects $X1$, $X2$, $X3$, $X4$, $X5$, and $X6$ in this order. Minimum number of misclassifications (MNM) of Revised IP-OLDF (Shinmura 1998, 2000a, b; Shinmura and Tarumi 2000) and NM of quadratic discriminant function (QDF) are zeroes in the one-variable model ($X1$). On the other hand, QDF misclassifies all regular cars as small cars after $X3$ enters the three-variable model because the number of seats of small cars is four. These data are more suitable for explaining Problem 3 than examination scores that use 100 items (Shinmura 2011b)

---

[1] See many studies of DEA analysis at Japanese researcher's DB: http://researchmap.jp/read0049917/. You can download over 14 papers from Misc(ellanies) category after 2013.

© Springer Science+Business Media Singapore 2016
S. Shinmura, *New Theory of Discriminant Analysis After R. Fisher*,
DOI 10.1007/978-981-10-2164-0_7

because the former are very direct and clear. Examination scores using four testlets in Chap. 5 are free from Problem 3.

3. Explanation of Matroska Feature-selection Method for Microarray Dataset (Method 2)

When we discriminate six microarray datasets (Jeffery et al. 2006) by eight LDFs, only Revised IP-OLDF can naturally select features and reduce the high-dimension gene space to the small gene subspace that is a linearly separable model (Shinmura 2015e–s). We call all linearly separable models as "Matroskas" in the Theory 2. We establish the Theory 2 and find that the datasets consist of several disjoint small gene spaces (Matroskas). We call these spaces, "the small Matroska (SMs)" with MNMs = 0. Because LSD discrimination is not popular now and Method 2 has several unknown ideas, we explain these ideas through these data in addition to Swiss banknote data (Flury and Rieduyl 1988).

If the data are LSD, the full model is the largest Matroska that contains many smaller Matroska in it. We already know that the smallest Matroska (the basic gene set, BGS) can describe the Matroska structure completely because MNM decreases monotonously. On the other hand, LASSO (Buhlmann and Geer 2011; Simon et al. 2013) attempts to make the feature-selection. If it cannot find SM or BGS in these data, it cannot explain the structure of Matroska in microarray datasets (Shinmura 2016b).

## 7.2 Japanese-Automobile Data

### 7.2.1 Data Outlook

Let us consider the discrimination of 29 regular and 15 small cars with the six variables listed in Table 7.1. Small cars have a unique Japanese specification. Women tend to buy them as second cars because they are cost efficient. The emission rate and capacity (number of seats) of small cars are smaller than regular cars. The range of the emission rate of small and regular cars is [0.657, 0.658] and [0.996, 3.456], respectively. We can discriminate the data by $X1 = (0.658 + 0.996)/2 = 0.827$. The number of seats for small and regular cars is four and [5, 8], respectively. We can discriminate the data by $X3 = (4 + 5)/2 = 4.5$. We understand that each $X1$ or $X3$ separates two classes completely and is BGS in Chap. 8. "$p$" is the number of variables selected by the forward stepwise procedure. Five- and six-variable models of Fisher's LDF (Fisher 1936, 1956) are zeroes. QDF and Revised IP-OLDF can find that the one-variable model ($X1$: emission) is a linearly separable model. The last two columns of Table 7.1 are NMs of regularized discriminant analysis (RDA) (Friedman 1989). Before 2012, JMP (Sall et al. 2004) switched QDF with RDA when QDF found data problems. However, both QDF

## 7.2 Japanese-Automobile Data

**Table 7.1** Comparison of MNM and NMs

| p | Var | t | LDF | QDF | MNM | $\lambda = \gamma = 0.8$ | $\lambda = \gamma = 0.1$ |
|---|---|---|---|---|---|---|---|
| 1 | Emission (X1) | 11.37 | 2 | 0 | 0 | 2 | 0 |
| 2 | Price (X2) | 5.42 | 1 | 0 | 0 | 4 | 0 |
| 3 | Capacity (X3) | 8.93 | 1 | 29 | 0 | 3 | 0 |
| 4 | $CO_2$ (X4) | 4.27 | 1 | 29 | 0 | 4 | 0 |
| 5 | Fuel (X5) | −4.00 | 0 | 29 | 0 | 5 | 0 |
| 6 | Sales (X6) | −0.82 | 0 | 29 | 0 | 5 | 0 |

and RDA misclassify all regular cars as small cars in the three-variable model (X1, X2, X3) because the number of seats of small cars is four (Problem 3). If we add slight random noise to the constant values, NMs become zero. Problem 3 might be the defect of the generalized inverse matrix technique incremented in QDF of JMP. After this fact, modified RDA was released. We must select two parameters $\lambda$ and $\gamma$. Therefore, we select the best combination, such as $\lambda = \gamma = 0.1$ with an 11 × 11 grid search. Although the best combination is valid for these data, we use these values for other data because we believe that the survey of the best combination is not meaningful for many users.

Figure 7.1 shows the box–whisker plots for the emission (X1) and capacity (X3). These graphs indicate that we can obtain two one-variable models that are linearly separable models and two BGSs. Therefore, 48 models, including (X1) and (X3) that are two BGSs, are linear separable among 63 models by monotonic decrease of MNM. The other 15 models are not linearly separable.

Table 7.2 lists the NMs of 48 linearly separable models. We categorize these models into three groups. The first 16 models, from SN = 1 to SN = 16, include both X1 and X3. The next 16 models, from SN = 17 to SN = 32, include only X1. The last 16 models, from SN = 33 to SN = 48, contain only X3. NMs of six MP-based LDFs, logistic regression (Cox 1958, Firth 1993), and RDA are zero. Therefore, these results are omitted from the table. On the other hand, 41 NMs of Fisher's LDF are not zero. Moreover, QDF misclassifies all regular cars as small

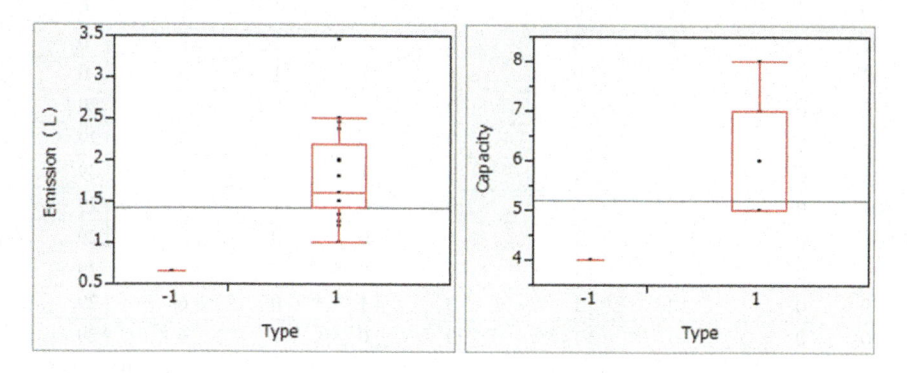

**Fig. 7.1** Box–whisker plots of emission and capacity (−1: small car, 1: regular car)

**Table 7.2** NMs of eight LDFs by forty-eight linear separable models

| SN | Emission | Price | Capacity | $CO_2$ | Fuel | Sales | LDF | QDF |
|---|---|---|---|---|---|---|---|---|
| 1 | 1 | 0 | 1 | 0 | 0 | 0 | 2 | 29 |
| 2 | 1 | 1 | 1 | 0 | 0 | 0 | 1 | 29 |
| 3 | 1 | 0 | 1 | 1 | 0 | 0 | 1 | 29 |
| 4 | 1 | 0 | 1 | 0 | 1 | 0 | 1 | 29 |
| 5 | 1 | 0 | 1 | 0 | 0 | 1 | 2 | 29 |
| 6 | 1 | 1 | 1 | 1 | 0 | 0 | 1 | 29 |
| 7 | 1 | 1 | 1 | 0 | 1 | 0 | 1 | 29 |
| 8 | 1 | 1 | 1 | 0 | 0 | 1 | 1 | 29 |
| 9 | 1 | 0 | 1 | 1 | 1 | 0 | 1 | 29 |
| 10 | 1 | 0 | 1 | 1 | 0 | 1 | 1 | 29 |
| 11 | 1 | 0 | 1 | 0 | 1 | 1 | 1 | 29 |
| 12 | 1 | 1 | 1 | 1 | 1 | 0 | 0 | 29 |
| 13 | 1 | 1 | 1 | 1 | 0 | 1 | 0 | 29 |
| 14 | 1 | 1 | 1 | 0 | 1 | 1 | 1 | 29 |
| 15 | 1 | 0 | 1 | 1 | 1 | 1 | 1 | 29 |
| 16 | 1 | 1 | 1 | 1 | 1 | 1 | 0 | 29 |
| 17 | 1 | 0 | 0 | 0 | 0 | 0 | 2 | 0 |
| 18 | 1 | 1 | 0 | 0 | 0 | 0 | 1 | 0 |
| 19 | 1 | 0 | 0 | 1 | 0 | 0 | 1 | 0 |
| 20 | 1 | 0 | 0 | 0 | 0 | 1 | 2 | 0 |
| 21 | 1 | 0 | 0 | 0 | 1 | 0 | 2 | 0 |
| 22 | 1 | 1 | 0 | 1 | 0 | 0 | 1 | 0 |
| 23 | 1 | 1 | 0 | 0 | 1 | 0 | 1 | 0 |
| 24 | 1 | 1 | 0 | 0 | 0 | 1 | 1 | 0 |
| 25 | 1 | 0 | 0 | 1 | 1 | 0 | 1 | 0 |
| 26 | 1 | 0 | 0 | 1 | 0 | 1 | 1 | 0 |
| 27 | 1 | 0 | 0 | 0 | 1 | 1 | 4 | 0 |
| 28 | 1 | 1 | 0 | 1 | 1 | 0 | 0 | 0 |
| 29 | 1 | 1 | 0 | 1 | 0 | 1 | 1 | 0 |
| 30 | 1 | 1 | 0 | 0 | 1 | 1 | 1 | 0 |
| 31 | 1 | 0 | 0 | 1 | 1 | 1 | 2 | 0 |
| 32 | 1 | 1 | 0 | 1 | 1 | 1 | 0 | 0 |
| 33 | 0 | 0 | 1 | 0 | 0 | 0 | 0 | 29 |
| 34 | 0 | 1 | 1 | 0 | 0 | 0 | 5 | 29 |
| 35 | 0 | 0 | 1 | 0 | 1 | 0 | 3 | 29 |
| 36 | 0 | 0 | 1 | 0 | 0 | 1 | 1 | 29 |
| 37 | 0 | 0 | 1 | 1 | 0 | 0 | 0 | 29 |
| 38 | 0 | 1 | 1 | 1 | 0 | 0 | 5 | 29 |
| 39 | 0 | 1 | 1 | 0 | 1 | 0 | 6 | 29 |
| 40 | 0 | 1 | 1 | 0 | 0 | 1 | 6 | 29 |

(continued)

## 7.2 Japanese-Automobile Data

**Table 7.2** (continued)

| SN | Emission | Price | Capacity | CO$_2$ | Fuel | Sales | LDF | QDF |
|---|---|---|---|---|---|---|---|---|
| 41 | 0 | 0 | 1 | 1 | 1 | 0 | 3 | 29 |
| 42 | 0 | 0 | 1 | 0 | 1 | 1 | 3 | 29 |
| 43 | 0 | 0 | 1 | 1 | 0 | 1 | 1 | 29 |
| 44 | 0 | 1 | 1 | 1 | 1 | 0 | 4 | 29 |
| 45 | 0 | 1 | 1 | 1 | 0 | 1 | 5 | 29 |
| 46 | 0 | 1 | 1 | 0 | 1 | 1 | 6 | 29 |
| 47 | 0 | 0 | 1 | 1 | 1 | 1 | 4 | 29 |
| 48 | 0 | 1 | 1 | 1 | 1 | 1 | 5 | 29 |

cars for the first and third groups because these models include X3. If we add slight random noise to X3 that belongs to small cars, all NMs become zero. Therefore, we conclude that only Fisher's LDF cannot discriminate 41 linearly separable models among 48 such models. Therefore, Fisher's LDF based on variance–covariance matrices is assumed to never discriminate LSD correctly and is unable to select features in gene analysis. Researchers of feature-selection method have better evaluate their method by common data before they try to evaluate their theory for the microarray datasets (the datasets). Although statisticians tried to discriminate the datasets by Fisher's LDF and QDF based on variance–covariance matrices over than ten years ago, their researches were failure because Fisher's LDF and QDF could not discriminate LSD correctly (Theory 2) (Shinmura 2014a, 2015b, d).

### 7.2.2 Comparison of Nine Discriminant Functions for Non-LSD

Table 7.3 lists NMs and number of cases on the discriminant hyperplane ($f(\mathbf{x}_i) = 0$) by 15 nonlinearly separable models (MNM > =1). Four variables correspond to four columns. We omit X1 and X3 from the table. The sixth column is MNM of Revised IP-OLDF. The next eight columns are "Diff1" of eight discriminant functions. LINGO (Schrage 2006) solves five MP-based LDFs, such as Revised IP-OLDF, Revised LP-OLDF (LP), Revised IPLP-OLDF (IPLP), SVM4, and SVM1 (Vapnik 1995). JMP solves four LDFs, such as logistic regression (Logistic), Fisher's LDF (LDF), QDF, and RDA. "Diff1" is the difference defined as (eight NMs – MNM). We omit NM of H-SVM because it cannot discriminate these models. The last column in the table is the number of cases on $f(\mathbf{x}_i) = 0$ of Revised LP-OLDF. We confirm that Revised LP-OLDF is not free from Problem 1. Because the other LDFs are free from Problem 1 in these data, we omit from the table seven columns for the number of cases on $f(\mathbf{x}_i) = 0$. The figures in bold are the maximum values among eight "Diff1s." There are many maximum values in LDF, QDF, and RDA compared with the other five LDFs. The last row is the number of figures in bold. It represents an approximate ranking of eight discriminant functions.

**Table 7.3** Diff1s and the number on $f(x) = 0$ by 15 linear separable models

| SN | Variable | | | | MNM | Diff1s | | | | | | | | $f(x_i)$ = 0 |
|---|---|---|---|---|---|---|---|---|---|---|---|---|---|---|
| | X2 | X4 | X5 | X6 | RIP | SVM4 | SVM1 | LP | IPLP | Logistic | LDF | QDF | RDA | LP |
| 49 | 1 | 0 | 0 | 0 | 5 | 1 | 1 | 1 | 0 | 0 | **8** | 2 | 7 | 0 |
| 50 | 0 | 0 | 1 | 0 | 10 | 1 | 1 | 1 | 0 | 1 | 1 | 2 | 2 | 1 |
| 51 | 0 | 1 | 0 | 0 | 10 | 1 | 1 | 1 | 0 | 1 | 3 | 2 | 2 | 1 |
| 52 | 0 | 0 | 0 | 1 | 13 | **2** | **2** | **2** | **2** | 1 | 2 | 1 | 2 | 0 |
| 53 | 1 | 0 | 1 | 0 | 4 | 2 | 2 | 2 | 0 | 1 | 4 | 3 | 6 | 0 |
| 54 | 1 | 1 | 0 | 0 | 4 | 2 | 2 | 2 | 0 | 2 | 7 | 4 | **8** | 0 |
| 55 | 0 | 0 | 1 | 1 | 8 | **6** | **6** | **6** | 1 | 3 | 3 | 1 | 1 | 0 |
| 56 | 1 | 0 | 0 | 1 | 4 | 2 | 2 | 2 | 0 | 1 | **8** | 3 | **8** | 0 |
| 57 | 0 | 1 | 1 | 0 | 10 | 2 | 2 | 1 | 0 | 2 | 1 | **4** | 1 | 1 |
| 58 | 0 | 1 | 0 | 1 | 8 | 3 | 3 | 3 | 0 | **4** | **5** | 4 | 4 | 0 |
| 59 | 1 | 0 | 1 | 1 | 4 | 2 | 2 | 2 | 0 | 1 | 4 | 3 | 6 | 0 |
| 60 | 1 | 1 | 1 | 0 | 4 | 0 | 0 | 0 | 0 | 0 | **5** | **5** | **9** | 0 |
| 61 | 1 | 1 | 0 | 1 | 4 | 2 | 2 | 2 | 0 | 1 | 7 | 4 | 7 | 0 |
| 62 | 0 | 1 | 1 | 1 | 8 | **7** | **7** | **7** | 0 | 3 | 2 | 2 | 3 | 0 |
| 63 | 1 | 1 | 1 | 1 | 3 | 1 | 1 | 1 | 0 | 1 | 6 | **5** | **9** | 0 |
| | Number of figures in **bold** | | | | | 3 | 3 | 3 | 1 | **0** | 5 | 2 | 9 | |

## 7.2.3 Consideration of Statistical Analysis

Figure 7.2 shows the score plots of PCA by six independent variables. The x-axis is the first principal component, and the y-axes are the second and third principal components. The left small 99 % probability ellipses are the small cars plotted with the symbol ".". The right large 99 % probability ellipses are the regular cars plotted with the symbol "+." Although these data are linearly separable, the regular car ellipses include small cars. Many researchers misunderstand PCA can detect LSD clearly.

If we use indicator $y_i$ as the dependent variable and analyze the Japanese-automobile data through regression analysis, the obtained regression

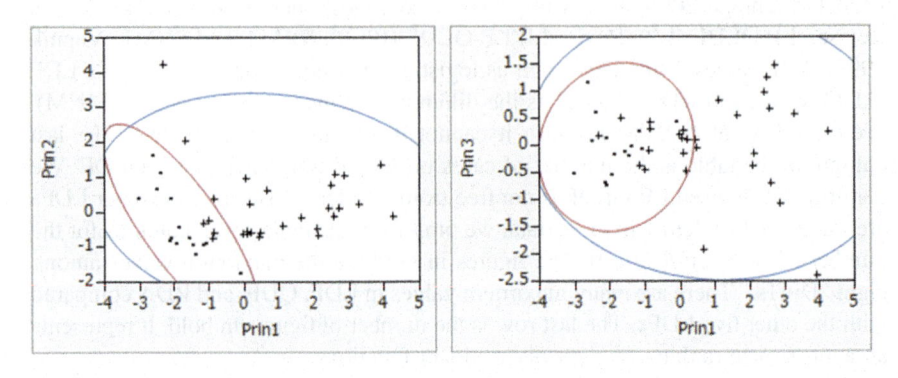

**Fig. 7.2** Score plots by PCA

## 7.2 Japanese-Automobile Data

coefficients are proportional to the discriminant coefficients by the plug-in rule1. Therefore, we can use the model selection procedures and statistics of the regression analysis instead of LOO procedure (Lachenbruch and Mickey 1968). Table 7.4 lists a summary of all possible combinations of six variables (Goodnight 1978). We sort six one-variable models in descending order by $R$-squares. After the two-variable model, the forward stepwise procedure selects these five models. AIC and BIC recommend the five-variable model, and Cp statistics suggests full mode. Because our research proposes a one-variable model ($X3$) in Tables 7.5, we must examine whether we can use these statistics for discriminant analysis in the near future. These statistics are independent of LSD discrimination.

**Table 7.4** Summary of all possible combinations of six variables

| p | Model | R-square | RMSE | AIC | BIC | Cp | RIP | LDF |
|---|---|---|---|---|---|---|---|---|
| 1 | X1 | 0.61 | 0.30 | 24.2 | 28.9 | 47.8 | 0 | 2 |
| 1 | X3 | 0.49 | 0.35 | 35.9 | 40.7 | 74.8 | 0 | 0 |
| 1 | X2 | 0.27 | 0.42 | 52.0 | 56.8 | 125.5 | 6 | 13 |
| 1 | X5 | 0.25 | 0.42 | 53.1 | 57.8 | 129.4 | 11 | 11 |
| 1 | X4 | 0.23 | 0.42 | 54.0 | 58.8 | 133.2 | 12 | 13 |
| 1 | X6 | 0.02 | 0.48 | 65.0 | 69.8 | 182.4 | 15 | 15 |
| 2 | X1, X2 | 0.77 | 0.24 | 4.3 | 10.4 | 14.9 | 0 | 1 |
| 3 | X1, X2, X3 | 0.79 | 0.23 | 1.7 | 9.0 | 11.0 | 0 | 1 |
| 4 | X1, X2, X3, X4 | 0.82 | 0.22 | −1.4 | 7.0 | 7.3 | 0 | 1 |
| 5 | X1, X2, X3, X4, X5 | 0.84 | 0.21 | **−3.4** | **6.0** | 5.0 | 0 | 0 |
| 6 | X1, X2, X3, X4, X5, X6 | 0.84 | 0.21 | −0.4 | 9.8 | **7.0** | 0 | 0 |

**Table 7.5** $M1$s and $M2$s of eight LDFs

| RIP | | M1 | M2 | Diff. | Model | | |
|---|---|---|---|---|---|---|---|
| 50s | 1 | 0 | 0 | 0 | X1, X2, X3 | | |
| | 2 | 0 | 0 | 0 | X1, X2 | | |
| | 3 | 0 | 0.07 | 0.07 | X1, X3 | | |
| | 4 | 0 | 0 | 0 | X2, X3 | | |
| | 5 | 0 | 0 | 0 | X1 | | |
| | 6 | 0 | 0 | 0 | X3 | | |
| | 7 | 9.55 | 12.75 | 3.2 | X2 | | |
| HSVM | | M1 | M2 | Diff. | M1Diff. | | M2Diff. |
| 38s | 1 | 0 | 0.11 | 0.11 | 0 | | 0.11 |
| | 2 | 0 | 0.2 | 0.2 | 0 | | 0.2 |
| | 3 | 0 | 0 | 0 | 0 | | −0.07 |
| | 4 | 0 | 0.11 | 0.11 | 0 | | 0.11 |
| | 5 | 0 | 0 | 0 | 0 | | 0 |
| | 6 | 0 | 0 | 0 | 0 | | 0 |

(continued)

**Table 7.5** (continued)

| SVM4 |   | $M1$ | $M2$ | Diff. | $M1$Diff. | $M2$Diff. |
|---|---|---|---|---|---|---|
| 40s | 1 | 0 | 0.11 | 0.11 | 0 | 0.11 |
|  | 2 | 0 | 0.2 | 0.2 | 0 | 0.2 |
|  | 3 | 0 | **0** | 0 | 0 | **−0.07** |
|  | 4 | 0 | 0.11 | 0.11 | 0 | 0.11 |
|  | 5 | 0 | **0** | 0 | 0 | **0** |
|  | 6 | 0 | **0** | 0 | 0 | **0** |
|  | 7 | 40.45 | 40.64 | 0.18 | 30.91 | **27.89** |
| SVM1 |   | $M1$ | $M2$ | Diff. | $M1$Diff. | $M2$Diff. |
| 44s | 1 | 1.14 | 1.2 | 0.07 | 1.14 | 1.2 |
|  | 2 | 0.98 | 1.7 | 0.73 | 0.98 | 1.7 |
|  | 3 | 0 | **0** | 0 | 0 | **−0.07** |
|  | 4 | 0.34 | 0.5 | 0.16 | 0.34 | 0.5 |
|  | 5 | 0.73 | **0.84** | 0.11 | 0.73 | **0.84** |
|  | 6 | 0 | **0** | 0 | 0 | **0** |
|  | 7 | 12.39 | 12.98 | 0.59 | 2.84 | 0.23 |
| LP | 1  32s | 0 | 0 | 0 | 0 | 0 |
|  | 2 | 0 | 0 | 0 | 0 | 0 |
|  | 3 | 0 | **0** | 0 | 0 | **−0.07** |
|  | 4 | 0 | 0 | 0 | 0 | 0 |
|  | 5 | 0 | **0** | 0 | 0 | **0** |
|  | 6 | 0 | **0** | 0 | 0 | **0** |
|  | 7 | 12.39 | 12.98 | 0.59 | 2.84 | 0.23 |
| IPLP | 1  1m40s | 0 | 0.27 | 0.27 | 0 | 0.27 |
|  | 2 | 0 | 0 | 0 | 0 | 0 |
|  | 3 | 0 | **0** | 0 | 0 | **−0.07** |
|  | 4 | 0 | 0 | 0 | 0 | 0 |
|  | 5 | 0 | **0** | 0 | 0 | **0** |
|  | 6 | 0 | **0** | 0 | 0 | **0** |
|  | 7 | 9.55 | 12.77 | 3.23 | 0 | 0.02 |
| Logistic | 1 | 0 | 0.36 | 0.36 | 0 | 0.36 |
|  | 2 | 0 | 0.05 | 0.05 | 0 | 0.05 |
|  | 3 | 0 | **0.02** | 0.02 | 0 | **−0.05** |
|  | 4 | 0 | 0 | 0 | 0 | 0 |
|  | 5 | 0 | **0** | 0 | 0 | **0** |
|  | 6 | 0 | **0** | 0 | 0 | **0** |
|  | 7 | 12.3 | 13.01 | 0.72 | 2.75 | 0.26 |
| LDF |   | $M1$ | $M2$ | Diff. | $M1$Diff. | $M2$Diff. |
|  | 1 | 1.5 | 2.35 | 0.85 | 1.5 | 2.35 |
|  | 2 | 1.89 | 2.91 | 1.03 | 1.89 | 2.91 |
|  | 3 | 4.52 | **4.75** | 0.23 | 4.52 | **4.68** |

(continued)

## 7.3 100-Fold Cross-Validation (Method 1)

**Table 7.5** (continued)

| | | | 2.4 | 10.43 | 12.83 |
|---|---|---|---|---|---|
| 4 | | | | | |
| 5 | 5.36 | **5.74** | 0.38 | 5.36 | **5.74** |
| 6 | 4.7 | **6.09** | 1.38 | 4.7 | **6.09** |
| 7 | 26.89 | 27.03 | 0.14 | 17.34 | **14.28** |

## 7.3 100-Fold Cross-Validation (Method 1)

### 7.3.1 Comparison of Best Model

In this section, we compare six MP-based LDFs, including H-SVM, and two statistical LDFs by Method 1. We examine the seven models made by three variables, such as $X1$, $X2$, and $X3$, because the other tree variables are not necessary for the discrimination explained by Table 7.1. Therefore, there are six linearly separable models, including $X1$ and $X3$, and one one-variable model ($X2$) that is not linearly separable. Table 7.5 lists the results of the Method 1. We omit QDF and RDA because they are not LDFs and have several defects. We examine the seven discriminant models of seven LDFs and six models of H-SVM. The first seven rows are the seven discriminant models of Revised IP-OLDF (RIP). The "Model" column shows the independent variable. $M1$s and $M2$s are the error rate means from the training and validation samples, respectively. Six $M1$s and $M2$s of Revised LP-OLDF are zeroes. Six $M1$s and five $M2$s of Revised IP-OLDF and Revised IPLP-OLDF are zeroes. Six $M1$s and three $M2$s of H-SVM, SVM4, and logistic regression are zeroes. Only two $M1$s and $M2$s of SVM1 are zeroes. All $M1$s and $M2$s of Fisher's LDF are not zeroes. We can summarize the results of the six linearly separable models as follows:

1. We can approximately evaluate the ranking of eight LDFs as follows: Revised LP-OLDF is the first rank. This result is entirely different from the other data and may be caused by Problem 1. Revised IP-OLDF and Revised IPLP-OLDF are the second grades. H-SVM, SVM4, and logistic regression are the third grades. Although SVM1 and Fisher's LDF are the fourth and fifth rank, these two LDFs cannot recognize the linearly separable models.
2. We can determine that the fifth and sixth models are the best models for two reasons: Although several models have "minimum $M2 = 0$," we select these models by the principal parsimony because those are a one-variable model. Revised IP-OLDF, SVM4, Revised LP-OLDF, Revised IPLP-OLDF, and logistic regression select these two models as the best models.

The seventh model is not linearly separable. The six values of "M2Diff" are 27.89, 0.23, 0.23, 0.02, 0.26, and 14.28 %. SVM4 and Fisher's LDF are 27.89 and 14.28 % worse, respectively, than Revised IP-OLDF. Although the discrimination of one variable is not important, we must investigate the reason that these two results are poor. On the other hand, the absolute values of other LDFs are within 0.26 %. In the pass/fail determinations (Shinmura 2015a, b), the other seven LDFs are similar to Revised IP-OLDF for LSD. The results listed in Table 7.5 are different from other data.

## 7.3.2  95 % CI of Coefficients by Six MP-Based LDFs

Tables 7.6 and 7.7 list the 95 % CI of six MP-based LDFs. Because we set all intercepts to one (Shinmura 2015b, 2016a), those are omitted from the tables. If 100 coefficients are constant, those are shown in "Median" rows.

### 7.3.2.1  Revised IP-OLDF Versus H-SVM

Equation (7.1) is the full model of Revised IP-OLDF (Shinmura 2003, 2004, 2005, 2007, 2010a, 2011a). Three coefficients are the constant, such as "0, 0, −0.2222…". This means that the discriminant hyperplanes are $X3 = 1/0.222 = 4.5$, which is

Table 7.6  95 % CI of eight LDFs

|   | RIP | $X1$ | $X2$ | $X3$ | H-SVM | $X1$ | $X2$ | $X3$ |
|---|---|---|---|---|---|---|---|---|
| 1 | 97.50 % |   |   | 0.975 |   | −0.071 | 4.50E−08 | −1.77E−01 |
|   | Median | 0 | 0 | −0.222 | Median | **−0.073** | 0 | **−0.209** |
|   | 2.50 % |   |   | 0.025 |   | −0.116 | −7.20E−08 | −0.218 |
| 2 | 97.50 % | −1.053 |   | 0.975 |   | −0.5115 | 9.62E−07 |   |
|   | Median | **-1.209** | 0 | Median |   | **−1.2092** | 0 |   |
|   | 2.50 % | −2.184 |   | 0.025 |   | −2.4382 | −3.50E−07 |   |
| 3 | 97.50 % | 379,174 |   | 0.975 |   | −0.0707 |   | −0.1978 |
|   | Median | −1.0776 | 0 | Median |   | **−0.0707** |   | **−0.2092** |
|   | 2.50 % | −1.2092 |   | 0.025 |   | −0.1155 |   | −0.2092 |
| 4 | 97.50 % |   |   | 0.975 |   |   | 0.00E+00 | −0.222 |
|   | Median |   | 0 | **−0.222** | Median |   | **0.00E+00** | **−0.222** |
|   | 2.50 % |   |   | 0.025 |   |   | 0.00E+00 | −0.222 |
| 5 | 97.50 % | −1.053 |   | 0.975 |   | −1.053 |   |   |
|   | Median | **−1.209** |   | Median |   | **−1.209** |   |   |
|   | 2.50 % | −1.209 |   | 0.025 |   | −1.209 |   |   |
| 6 | 97.50 % |   |   | 0.975 |   |   |   |   |
|   | Median |   | −0.222 | Median |   |   |   | −0.222 |
|   | 2.50 % |   |   | 0.025 |   |   |   |   |
| 7 | 97.50 % |   | −6.70E−07 |   |   |   |   |   |
|   | Median |   | **−7.50E−07** |   |   |   |   |   |
|   | 2.50 % |   | −8.20E−07 |   |   |   |   |   |

7.3 100-Fold Cross-Validation (Method 1)

**Table 7.7** 95% CI of four LDFs

| LP | X1 | X2 | X3 | IPLP | X1 | X2 | X3 | SVM4 | X1 | X2 | X3 | SVM1 | X1 | X2 | X3 |
|---|---|---|---|---|---|---|---|---|---|---|---|---|---|---|---|
| 97.50 % | 0 | 0.00E+00 | −0.222 | 97.50 % | 3.194 | 1.38E−06 | 0.19968 | 97.50 % | −0.071 | 4.13E−08 | −0.172 | 97.50 % | −0.0403 | 4.25E−09 | −8.89E−02 |
| Median | **0** | **0** | **−0.222** | Median | 0 | 0 | −0.2222 | Median | **−0.0806** | 9.60E−09 | **−0.2101** | Median | **−0.0876** | 0 | **−0.2037** |
| 2.50 % | 0 | 0.00E+00 | −0.222 | 2.50 % | −3.966 | −0.00000084 | −0.6627 | 2.50 % | −0.1196 | −8.50E−08 | −0.216 | 2.50 % | −0.1216 | −9.20E−08 | −0.2177 |
| 97.50 % | −1.0526 | 9.36E−07 | | 97.50 % | −1.0526 | 9.36E−07 | | 97.50 % | −0.5118 | 9.22E−07 | | 97.50 % | −0.1276 | 1.05E−07 | |
| Median | **−1.2092** | **0** | | Median | **−2.0473** | **7.11E−07** | | Median | **−1.2101** | 5.34E−09 | | Median | **−1.0055** | 0 | |
| 2.50 % | −2.3857 | 0.00E+00 | | 2.50 % | −2.3857 | 0 | | 2.50 % | −2.3684 | −7.40E−07 | | 2.50 % | −1.1712 | −6.50E−07 | |
| 97.50 % | 0 | | −0.222 | 97.50 % | 0 | | −0.222 | 97.50 % | −0.0707 | | −0.1978 | 97.50 % | −0.0707 | | −0.197 |
| Median | **0** | | **−0.222** | Median | **0** | | **−0.222** | Median | **−0.0707** | | **−0.2092** | Median | **−0.0876** | | **−0.2037** |
| 2.50 % | 0 | | −0.222 | 2.50 % | 0 | | −0.222 | 2.50 % | −0.1155 | | −0.2092 | 2.50 % | −0.1155 | | −0.2092 |
| 97.50 % | | 0 | −0.222 | 97.50 % | | 0 | −0.222 | 97.50 % | | −1.00E−10 | −0.2152 | 97.50 % | | 2.15E−09 | −0.0695 |
| Median | | **0** | **−0.222** | Median | | **0** | **−0.222** | Median | | **−1.50E−09** | **−0.2218** | Median | | 0.00E+00 | **−0.2222** |
| 2.50 % | | 0 | −0.222 | 2.50 % | | 0 | −0.222 | 2.50 % | | −2.40E−07 | −0.222 | 2.50 % | | −1.40E−07 | −0.2238 |
| 97.50 % | −1.053 | | | 97.50 % | −1.053 | | | 97.50 % | −1.0526 | | | 97.50 % | −0.9285 | | |
| Median | **−1.209** | | | Median | **−1.209** | | | Median | **−1.2092** | | | Median | **−1.0065** | | |
| 2.50 % | −1.209 | | | 2.50 % | −1.209 | | | 2.50 % | −1.2092 | | | 2.50 % | −1.0526 | | |
| 97.50 % | | | −0.222 | 97.50 % | | | −0.222 | 97.50 % | | | −0.222 | 97.50 % | | | −0.222 |
| Median | | | **−0.222** | Median | | | **−0.222** | Median | | | **−0.222** | Median | | | **−0.222** |
| 2.50 % | | | −0.222 | 2.50 % | | | −0.222 | 2.50 % | | | −0.222 | 2.50 % | | | −0.222 |
| 97.50% | | −0.0000007 | | 97.50 % | | −6.50E−07 | | 97.50 % | | 1.22E−05 | | 97.50 % | | −7.00E−07 | |
| Median | | **−0.00000074** | | Median | | **−7.50E−07** | | Median | | −7.20E−07 | | Median | | **−7.40E−07** | |
| 2.50 % | | −0.00000079 | | 2.50 % | | −8.10E−07 | | 2.50 % | | −1.40E−05 | | 2.50 % | | −7.90E−07 | |

similar to the discriminant hyperplane described in Sect. 7.2.1. Because the Japanese-automobile data are LSD and the one-variable model ($X3$) is a linearly separable model, 100 Revised IP-OLDFs select an intermediate point as the same discriminant hyperplane. On the other hand, three 95 % CIs of H-SVM are [−0.116, −0.071], [−7.2E−8, 4.5E−8], and [−0.218, −0.177]. Because H-SVM maximizes the distance of two SVs and 100 sets of two SVs are different, we estimate that H-SVM cannot select the features naturally. Because both $X1$ and $X2$ coefficients of RIP123 are zeroes, Method 2 suggests selecting $X3$ as the first BGS instead of $X1$. Although we select 5th and 6th models by the minimum $M2$ standard in Table 7.5, we can determine that the best model is $X3$ by Table 7.6. Moreover, the forward stepwise procedure and $t$ test initially select $X1$ in Table 7.1. We claim that the statistical suggestions are different from the 95 % CI of coefficients by Method 1. Equation (7.1) means that the discriminant hyperplane is $X3 = 4.5$. If $X3 < 4.5$, we can determine that the car belongs to the small car group. Otherwise, if $X3 > 4.5$, we can determine that the car belongs to the regular car group. This discriminant rule is similar to the fact that small cars have four seats, and regular cars have over five seats. This fact should be emphasized before the discrimination. However, we cannot perform the same action for more than two variables without Revised IP-OLDF. Moreover, we can understand that Revised IP-OLDF can select the features suited to human sensibility.

$$\text{RIP123} : 0 \times X1 + 0 \times X2 - 0.2222 \times X3 + 1 = -0.2222 \times X3 + 1 = 0 \quad (7.1)$$

Equation (7.2) is a two-variable model, such as ($X2$, $X3$). The 95 % CI of $X2$ and $X3$ is the same in the full model and one-variable model ($X3$). When we make the feature-selection by Revised IP-OLDF for 100 training samples, 100 Revised IP-OLDFs reduce the full models to 100 one-variable models ($X3$). Although Theory 1 evaluates 100 Revised IP-OLDFs, it can make the feature-selection naturally for two models, such as ($X2$, $X3$) and ($X3$).

$$\text{RIP23} : 0 \times X2 - 0.2222 \times X3 + 1 = -0.2222 \times X3 + 1 = 0 \quad (7.2)$$

Equation (7.3) is a two-variable model using the median, such as ($X1$, $X2$). We can determine that this model is linearly separable by $X1$ without $X2$. The discriminant hyperplane is $X1 = 1/1.209 = 0.827$, which is equal to an intermediate point, such as $X1 = (0.658 + 0.996)/2$, as described in Sect. 7.2.1. However, 100 intermediate points are different. The 95% CI of $X1$ in the one-variable model ($X1$) has almost the same 95 % CI of model ($X1$, $X2$). It reduce the two-variable model ($X1$, $X2$) to the fifth model ($X1$).

$$\text{RIP12} : -1.209^* \times X1 + 0 \times X2 + 1 = -1.209^* \times X1 + 1 = 0 \quad (7.3)$$

Equation (7.4) is a two-variable model, such as ($X1$, $X3$). Although each variable is a linearly separable model, both coefficients of ($X1$, $X3$) are zero at the 5 % level.

## 7.3 100-Fold Cross-Validation (Method 1)

We estimate that each variable disturbs the capability of feature-selection by another variable. Revised IP-OLDF cannot select the features by this model.

$$\text{RIP13} : -1.078 \times X1 + 0 \times X3 = 0 \tag{7.4}$$

Equation (7.5) is a one-variable model, such as (X2). The discriminant hyperplane is X2 = 1,333,333. The price of four regular cars is less than 1.3 million yen, and those of two small cars are higher than this price. Therefore, the error rate is 0.135 (= 6/44) (Miyake and Shinmura 1976).

$$\text{RIP2} : -0.00000075^{*} \times X2 + 1 = 0 \tag{7.5}$$

If we compare H-SVM with Revised IP-OLDF, two models, such as (X1), and (X3), are the same. Those are two BGSs.

### 7.3.2.2 Revised IPLP-OLDF, Revised LP-OLDF, and other LDFs

Four models, such as (X1, X2, X3), (X2, X3), (X1), and (X3) of Revised LP-OLDF, are similar to Eq. (7.2). Although Revised LP-OLDF is weak for Problem 1, it might be more suitable for LSD discrimination than H-SVM. However, there is no theoretical reason for it to discriminate LSD. On the other hand, the full model of Revised IPLP-OLDF (Shinmura 2010b, 2014b), SVM4, and SVM1 are not useful for feature-selection.

### *7.3.3 95 % CI of Coefficients by Fisher's LDF and Logistic Regression*

Table 7.8 lists the 95 % CI of Fisher's LDF and logistic regression. We assume that $y_i$ is the object variable and analyze the data by regression analysis (plug-in rule1). The first row is the coefficients. The second row is SE. The third row is the *p*-value instead of *t*-value by *t* test. We know that the only coefficient of X2 in the fourth model (X2, X3) of LDF is zero. The other coefficients are rejected at the 5 or 1 % levels. Therefore, it is difficult for us to select a good model among the six models. Because the seventh model (X2) is not linearly separable, this model is rejected at the 1 % level. We should not trust the SE and *p*-value of Fisher's LDF by the plug-in rule1. In addition to this recommendation, we cannot use the SE and p-value of logistic regression.

**Table 7.8** 95 % CI of Fisher's LDF and logistic regression

| LDF | X1 | X2 | X3 | c | Logistic | X1 | X2 | X3 | c |
|---|---|---|---|---|---|---|---|---|---|
| Coeff. | **1.84** | **−7.90E−07** | **0.179** | **−1.709** | Coeff. | **−19.488** | 6.93E−06 | −5.37E+01 | 251.697 |
| SE | 0.25 | 1.56E−07 | 0.08 | 0.32 | SE | 0 | 2.317 | 2979878 | 1283091 |
| P | 0.0001 | 0.0001 | 0.031 | 0.0001 | P | | 1 | 1 | 1 |
| Coeff. | **2.119** | **−8.41E−07** | | **−1.085** | Coeff. | **−179.039** | 1.75E−03 | | 87.858 |
| SE | 0.227 | 1.62E−07 | | 0.164 | SE | 0 | 1.774 | | 2497413 |
| P | 0.0001 | 0.0001 | | 0.0001 | P | | 1 | | 1 |
| Coeff. | **0.78** | | **0.235** | **−2.003** | Coeff. | −3.777 | | −34.414 | 158.844 |
| SE | 0.175 | | 0.101 | 0.399 | SE | 11,909 | | 8185.144 | 27565 |
| P | 0.0001 | | 0.024 | 0.0001 | P | 0.9997 | | 0.9966 | 0.9954 |
| Coeff. | | 1.62E−07 | **0.474** | **−2.443** | Coeff. | | −2.77E−07 | −62.889 | 283.495 |
| SE | | 1.31E−07 | 0.105 | 0.46 | SE | | 2.12E+00 | 2721988 | 11568553 |
| P | | 0.222 | 0.0001 | 0.0001 | P | | 1 | 1 | 1 |
| Coeff. | **1.068** | | | **−1.191** | Coeff. | −94.471 | | | 79.66 |
| SE | 0.131 | | | 0.207 | SE | 6289 | | | 5492.898 |
| P | 0.0001 | | | 0.0001 | P | 0.988 | | | 0.988 |
| Coeff. | | | **0.549** | **−2.528** | Coeff. | | | −37.05 | 166.806 |
| SE | | | 0.086 | 0.458 | SE | | | 3702.95 | 17078 |
| P | | | 0.0001 | 0.0001 | P | | | 0.992 | 0.992 |
| Coeff. | | **5.04E+00** | | −0.636 | Coeff. | | −6.97E−06 | | **9.233** |
| SE | | 1.29E−07 | | 0.273 | SE | | 2.60E−06 | | 3.421 |
| P | | 3.00E−04 | | 0.025 | P | | 7.30E−03 | | 0.007 |

## 7.4 Matroska Feature-Selection Method (Method 2)

### 7.4.1 Feature-Selection by Revised IP-OLDF

Table 7.9 lists 63 models. Because the first 32 models include $X1$, MNMs are zero. A total of 28 coefficients of $X1$ and intercept are 5.92 and −4.893, respectively. The discriminant hyperplane is $X1 = 4.893/5.92 = 0.82652$, similar to that described in Sect. 7.2.1. On the other hand, because the other four coefficients of $X3$ and intercept are 2 and −9, respectively, the discriminant hyperplane is $X3 = 9/2 = 4.5$, similar to that described in Sect. 7.2.1. The next 16 models are the same as the aforementioned four models. The last 15 models do not include $X1$ and $X3$; MNMs are not zero. We simulate Method 2 by this table. When we discriminate the full model (SN = 1), the only coefficient of $X1$ is 5.917, and the other five coefficients are zero. Therefore, we can reduce the six-dimension space to one-dimension subspace by natural feature-selection. We call the linearly separable model, "Matroska" in *Theory 2*. The large Matroska with six variables includes five five-variable models (or smaller Matroska) in it. Each five-variable model includes four four-variable models in it. Each four-variable model includes three three-variable models in it. Each three-variable model includes two two-variable models in it. The two two-variable models include the smallest linearly separable model ($X1$) in it. We call this model, "BGS." Therefore, there are 120 (=5 × 4 × 3 × 2) Matroska products, including $X1$.

## 7.4 Matroska Feature-Selection Method (Method 2)

**Table 7.9** 63 models by Revised IP-OLDF

| SN2 | Var. | RIP | x1 | x2 | x3 | x4 | x5 | x6 | C |
|---|---|---|---|---|---|---|---|---|---|
| 1 | 1–6 | 0 | 5.92 | 0 | 0 | 0 | 0 | 0 | −4.893 |
| 2 | 1–5 | 0 | 5.92 | 0 | 0 | 0 | 0 |  | −4.893 |
| 3 | 1–4, 6 | 0 | 5.92 | 0 | 0 | 0 |  | 0 | −4.893 |
| 4 | 1–3, 5, 6 | 0 | 5.92 | 0 | 0 |  | 0 | 0 | −4.893 |
| 5 | 1, 3–6 | 0 | 5.92 |  | 0 | 0 | 0 | 0 | −4.893 |
| 6 | 1, 2, 4–6 | 0 | 5.92 | 0 |  | 0 | 0 | 0 | −4.893 |
| 7 | 1–4 | 0 | 5.92 | 0 | 0 | 0 |  |  | −4.893 |
| 8 | 1–3, 5 | 0 | 5.92 | 0 | 0 |  | 0 |  | −4.893 |
| 9 | 1–3, 6 | 0 | 5.92 | 0 | 0 |  |  | 0 | −4.893 |
| 10 | 1, 3–5 | 0 | 5.92 |  | 0 | 0 | 0 |  | −4.893 |
| 11 | 1, 3, 4, 6 | 0 | 5.92 |  | 0 | 0 |  | 0 | −4.893 |
| 12 | 1, 3, 5, 6 | 0 | 5.92 |  | 0 |  | 0 | 0 | −4.893 |
| 13 | 1, 2, 4, 5 | 0 | 5.92 | 0 |  | 0 | 0 |  | −4.893 |
| 14 | 1, 2, 4, 6 | 0 | 5.92 | 0 |  | 0 |  | 0 | −4.893 |
| 15 | 1, 2, 5, 6 | 0 | 5.92 | 0 |  |  | 0 | 0 | −4.893 |
| 16 | 1, 4–6 | 0 | 5.92 |  |  | 0 | 0 | 0 | −4.893 |
| 17 | 1–3 | 0 | 0 | 0 | 2 |  |  |  | −9 |
| 18 | 1, 3, 4 | 0 | 0 |  | 2 | 0 |  |  | −9 |
| 19 | 1, 3, 5 | 0 | 0 |  | 2 |  | 0 |  | −9 |
| 20 | 1, 3, 6 | 0 | 5.92 |  | 0 |  |  | 0 | −4.893 |
| 21 | 1, 2, 4 | 0 | 5.92 | 0 |  | 0 |  |  | −4.893 |
| 22 | 1, 2, 5 | 0 | 5.92 | 0 |  |  | 0 |  | −4.893 |
| 23 | 1, 2, 6 | 0 | 5.92 | 0 |  |  |  | 0 | −4.893 |
| 24 | 1, 4, 5 | 0 | 5.92 |  |  | 0 | 0 |  | −4.893 |
| 25 | 1, 4, 6 | 0 | 5.92 |  |  | 0 |  | 0 | −4.893 |
| 26 | 1, 5, 6 | 0 | 5.92 |  |  |  | 0 | 0 | −4.893 |
| 27 | 1, 3 | 0 | 0 |  | 2 |  |  |  | −9 |
| 28 | 1, 2 | 0 | 5.92 | 0 |  |  |  |  | −4.893 |
| 29 | 1, 4 | 0 | 5.92 |  |  | 0 |  |  | −4.893 |
| 30 | 1, 6 | 0 | 5.92 |  |  |  |  | 0 | −4.893 |
| 31 | 1, 5 | 0 | 5.92 |  |  |  | 0 |  | −4.893 |
| 32 | 1 | 0 | 5.92 |  |  |  |  |  | −4.893 |
| 33 | 2–6 | 0 |  | 0 | 2 | 0 | 0 | 0 | −9 |
| 34 | 2–4 | 0 |  | 0 | 2 | 0 |  |  | −9 |
| 35 | 2–4, 6 | 0 |  | 0 | 2 | 0 |  | 0 | −9 |
| 36 | 2, 3, 5, 6 | 0 |  | 0 | 2 |  | 0 | 0 | −9 |
| 37 | 3–6 | 0 |  |  | 2 | 0 | 0 | 0 | −9 |
| 38 | 2–4 | 0 |  | 0 | 2 | 0 |  |  | −9 |
| 39 | 2, 3, 5 | 0 |  | 0 | 2 |  | 0 |  | −9 |
| 40 | 2, 3, 6 | 0 |  | 0 | 2 |  |  | 0 | −9 |

(continued)

**Table 7.9** (continued)

| SN2 | Var. | RIP | x1 | x2 | x3 | x4 | x5 | x6 | C |
|---|---|---|---|---|---|---|---|---|---|
| 41 | 3–5 | 0 | | | 2 | 0 | 0 | | −9 |
| 42 | 3, 5, 6 | 0 | | | 2 | | 0 | 0 | −9 |
| 43 | 3, 4, 6 | 0 | | | 2 | 0 | | 0 | −9 |
| 44 | 2, 3 | 0 | | 0 | 2 | | | | −9 |
| 45 | 3, 5 | 0 | | | 2 | | 0 | | −9 |
| 46 | 3, 6 | 0 | | | 2 | | | 0 | −9 |
| 47 | 3, 4 | 0 | | | 2 | 0 | | | −9 |
| 48 | 3 | 0 | | | 2 | | | | −9 |
| 49 | 2 | 5 | | 0 | | | | | −134.3 |
| 50 | 2, 4–6 | 3 | | 0 | | −46 | −199 | 0 | 5342.8 |
| 51 | 2, 5, 6 | 4 | | 0 | | | 4.03 | 0 | −782.7 |
| 52 | 2, 4, 5 | 4 | | 0 | | −0.3 | −1.7 | | 45.1 |
| 53 | 2, 4, 6 | 4 | | 0.03 | | −121 | | −0.6 | −28,515 |
| 54 | 4–6 | 8 | | | | −96 | −809 | −0.2 | 29,809 |
| 55 | 2, 5 | 4 | | 0.03 | | | 45.4 | | −40,747 |
| 56 | 2, 4 | 4 | | 0 | | −0.1 | | | −461.8 |
| 57 | 5, 6 | 8 | | | | | −685 | −0.2 | 17,748 |
| 58 | 2, 6 | 4 | | 0.03 | | | | 0.09 | −40,125 |
| 59 | 4, 5 | 10 | | | | 3.54 | 10.8 | | −601.3 |
| 60 | 4, 6 | 8 | | | | 160 | | −0.3 | −14,026 |
| 61 | 5 | 10 | | | | | −2.5 | | 59.5 |
| 62 | 4 | 10 | | | | 90.7 | | | −8980 |
| 63 | 6 | 13 | | | | | | −0.7 | 6773.5 |

Next, when we remove $X1$ from the full model and discriminate five-variable model ($X2$–$X6$) by Revised IP-OLDF (SN2 = 33), the only coefficient of $X3$ is two and the other four coefficients are zeroes. Therefore, we can reduce the five-dimension space to one-dimension subspace. The second large Matroska includes four four-variable models in it. Each four-variable model includes three three-variable models in it. Each three-variable model includes two two-variable models in it. Each two-variable model includes BGS ($X3$) in it. Model $X3$ is the second BGS. Therefore, there are 24 (=4 × 3 × 2) Matroska products, including $X3$. Next, when we remove $X3$ from the five-variable model and discriminate the four-variable model ($X2$, $X4$–$X6$) by Revised IP-OLDF (SN2 = 50), two coefficients are zeroes. Because NM is not zero, we stop Method 2. Therefore, we can understand that the structure of these data consist of two BGSs, such as $X1$ and $X3$, and the other four-variable models that are not linearly separable. Many statisticians have struggled over ten years to analyze the high-dimension gene space. However, we should analyze two one-variable models because these two BGSs explain the Matroska structure easily. Although

## 7.4 Matroska Feature-Selection Method (Method 2)

Method 2 initially selects $X3$ for the three-variable model in Table 7.6, Method 2 initially selects $X1$ for the six-variable model in Table 7.9.

### 7.4.2 Coefficient of H-SVM and SVM4

Table 7.10 lists the coefficients of H-SVM. Although the three coefficients of $X2$, $X4$, and $X6$ are small in the full model (SN = 1), all coefficients of H-SVM are not zero. These results imply the following:

1. H-SVM cannot naturally select features. However, because the three coefficients of $X2$, $X4$, and $X6$ are minuscule, the coefficients might be zero by LASSO (Simon et al. 2013).
2. Although the RIP coefficients of $X1$ are 5.92/zero for 32 linearly separable models in Table 7.9, the coefficient of $X1$ for H-SVM varies in Table 7.10. If the $X3$ values for small cars vary, the $X3$ coefficient for H-SVM can also vary.

Table 7.11 lists the coefficients of SVM4. Although the three coefficients of $X2$, $X4$, and $X6$ are minuscule in the full model (SN2 = 1), all coefficients of SVM4 are not zero. SVM4 cannot make the feature-selection naturally.

**Table 7.10** Coefficients of H-SVM

| SN2 | x1 | x2 | x3 | x4 | x5 | x6 | x7 |
|---|---|---|---|---|---|---|---|
| 1 | 0.63 | −2.0E−07 | 1.78 | −9.0E−03 | −0.1 | 4.0E−06 | −6.1 |
| 2 | 0.62 | −2.0E−07 | 1.79 | −4.0E−03 | 0 | | −7.1 |
| 3 | 0.61 | −2.0E−07 | 1.79 | 2.0E−03 | | 1.0E−06 | −8.6 |
| 4 | 0.62 | −2.0E−07 | 1.78 | | 0 | 2.0E−06 | −8.1 |
| 5 | 0.61 | | 1.79 | −2.0E−04 | 0 | 4.0E−08 | −8.5 |
| 6 | 5.62 | −1.0E−06 | | −8.0E−02 | −0.5 | 3.0E−05 | 17.1 |
| 7 | 0.61 | 0 | 1.79 | 0 | | | −8.6 |
| 8 | 0.61 | −2.0E−07 | 1.79 | | 0 | | −8.1 |
| 9 | 0.61 | 0 | 1.79 | | | 0 | −8.6 |
| 10 | 0.61 | | 1.79 | −5.0E−08 | 0 | | −8.6 |
| 11 | 0.61 | | 1.79 | 0 | | 0 | −8.6 |
| 12 | 0.61 | | 1.79 | | 0 | 0 | −8.6 |
| 13 | 5.77 | −1.0E−06 | | −4.0E−02 | −0.4 | | 8.94 |
| 14 | 5.83 | −2.0E−06 | | 2.0E−02 | | 1.0E−05 | −4.7 |
| 15 | 5.74 | −2.0E−06 | | | −0.1 | 2.0E−05 | −0.3 |
| 16 | 5.92 | 0 | | | 0 | 0 | −4.9 |
| 17 | 0.61 | 0 | 1.79 | | | | −8.6 |

(continued)

**Table 7.10** (continued)

| SN2 | x1 | x2 | x3 | x4 | x5 | x6 | x7 |
|---|---|---|---|---|---|---|---|
| 18 | 0.61 |  | 1.79 | 0 | 0 | 0 | −8.6 |
| 19 | 0.61 | 0 | 1.79 | 0 |  |  | −8.6 |
| 20 | 0.61 |  | 1.79 |  | 0 |  | −8.6 |
| 21 | 5.9 | −8.0E−08 |  | 6.0E−04 |  |  | −4.9 |
| 22 | 5.86 | −2.0E−06 |  |  | −0.1 |  | −0.2 |
| 23 | 5.92 | 0 |  |  |  | 0 | −4.9 |
| 24 | 5.92 |  |  | 0 | 0 |  | −4.9 |
| 25 | 5.92 |  |  | 0 |  | 0 | −4.9 |
| 26 | 5.92 |  |  |  | 0 | 0 | −4.9 |
| 27 | 0.61 |  | 1.79 |  |  |  | −8.6 |
| 28 | 5.92 | 0 |  |  |  |  | −4.9 |
| 29 | 5.92 |  |  | 0 |  |  | −4.9 |
| 30 | 5.92 |  |  |  |  | 0 | −4.9 |
| 31 | 5.92 |  |  |  | 0 |  | −4.9 |
| 32 | 5.92 |  |  |  |  |  | −4.9 |
| 33 |  | 0 | 2 | 0 | 0 | 0 | −9 |
| 34 |  | 0 | 2 | 0 |  |  | −9 |
| 35 |  | 0 | 2 | 0 |  | 0 | −9 |
| 36 |  | 0 | 2 |  | 0 | 0 | −9 |
| 37 |  |  | 2 | 0 | 0 | 0 | −9 |
| 38 |  | 0 | 2 | 0 |  |  | −9 |
| 39 |  | 0 | 2 |  | 0 |  | −9 |
| 40 |  | 0 | 2 |  |  | 0 | −9 |
| 41 |  |  | 2 | 0 | 0 |  | −9 |
| 42 |  |  | 2 |  | 0 | 0 | −9 |
| 43 |  |  | 2 | 0 |  | 0 | −9 |
| 44 |  | 0 | 2 |  |  |  | −9 |
| 45 |  |  | 2 |  | 0 |  | −9 |
| 46 |  |  | 2 |  |  | 0 | −9 |
| 47 |  |  | 2 | 0 |  |  | −9 |
| 48 |  |  | 2 |  |  |  | −9 |

7.4 Matroska Feature-Selection Method (Method 2)

**Table 7.11** Coefficients of SVM4

| SN2 | $x1$ | $x2$ | $x3$ | $x4$ | $x5$ | $x6$ | $x7$ |
|---|---|---|---|---|---|---|---|
| 1 | 0.87 | −2.00E−08 | 1.74 | −1.00E−02 | −0.1 | 3.00E−06 | −5.1 |
| 2 | 0.87 | −4.00E−08 | 1.73 | −1.00E−02 | −0.1 | | −5.3 |
| 3 | 0.89 | −7.00E−08 | 1.8 | 2.00E−03 | | −3.00E−06 | −9 |
| 4 | 0.86 | −1.00E−07 | 1.77 | | 0 | 7.00E−07 | −8.3 |
| 5 | 0.61 | | 1.79 | −3.00E−03 | 0 | 5.00E−07 | −7.9 |
| 6 | 5.62 | −1.00E−06 | | −8.00E−02 | −0.5 | 3.00E−05 | 17.1 |
| 7 | 0.82 | −2.00E−07 | 1.73 | 2.00E−03 | | | −8.4 |
| 8 | 0.83 | −2.00E−07 | 1.73 | | 0 | | −8 |
| 9 | 0.81 | −2.00E−07 | 1.74 | | | −6.00E−06 | −8.3 |
| 10 | 0.61 | | 1.79 | −4.00E−06 | 0 | | −8.6 |
| 11 | 0.61 | | 1.79 | 0 | | 0 | −8.6 |
| 12 | 0.61 | | 1.79 | | 0 | −5.00E−09 | −8.6 |
| 13 | 5.77 | −1.00E−06 | | −4.00E−02 | −0.4 | | 8.95 |
| 14 | 5.83 | −2.00E−06 | | 2.00E−02 | | 1.00E−05 | −4.7 |
| 15 | 5.74 | −2.00E−06 | | | −0.1 | 2.00E−05 | −0.3 |
| 16 | 5.92 | | | −4.00E−04 | 0 | 7.00E−08 | −4.8 |
| 17 | 0.8 | −2.00E−07 | 1.74 | | | | −8.3 |
| 18 | 0.61 | | 1.79 | 0 | | | −8.6 |
| 19 | 0.61 | | 1.79 | | 0 | | −8.6 |
| 20 | 0.61 | | 1.79 | | | −2.00E−08 | −8.6 |
| 21 | 5.9 | −1.00E−07 | | 9.00E−04 | | | −4.9 |
| 22 | 5.86 | −2.00E−06 | | | −0.1 | | −0.2 |
| 23 | 5.91 | −8.00E−08 | | | | −6.00E−07 | −4.8 |
| 24 | 5.92 | | | −2.00E−06 | 0 | | −4.9 |
| 25 | 5.92 | | | 0 | | 0 | −4.9 |
| 26 | 5.92 | | | | 0 | 0 | −4.9 |
| 27 | 0.61 | | 1.79 | | | | −8.6 |
| 28 | 5.92 | 0 | | | | | −4.9 |
| 29 | 5.92 | | | 0 | | | −4.9 |
| 30 | 5.92 | | | | | 0 | −4.9 |
| 31 | 5.92 | | | | 0 | | −4.9 |
| 32 | 5.92 | | | | | | −4.9 |
| 33 | | 0 | 2 | −7.00E−08 | 0 | 0 | −9 |
| 34 | | 4.00E−09 | 2 | −9.00E−05 | | | −9 |
| 35 | | 2.00E−09 | 2 | 4.00E−06 | | −4.00E−08 | −9 |
| 36 | | 2.00E−07 | 2.1 | | 0 | −2.00E−06 | −9.6 |
| 37 | | | 2 | 7.00E−09 | 0 | 0 | −9 |
| 38 | | 1.00E−08 | 2.01 | 3.00E−05 | | | −9 |
| 39 | | 1.00E−08 | 2.01 | | 0 | | −9 |
| 40 | | 3.00E−07 | 2.17 | | | −6.00E−06 | −10 |

(continued)

**Table 7.11** (continued)

| SN2 | x1 | x2 | x3 | x4 | x5 | x6 | x7 |
|---|---|---|---|---|---|---|---|
| 41 | | | 2 | 3.00E−07 | 0 | | −9 |
| 42 | | | 2 | | 0 | 0 | −9 |
| 43 | | | 2 | 4.00E−07 | | 0 | −9 |
| 44 | | 4.00E−07 | 2.23 | | | | −11 |
| 45 | | | 2 | | 0 | | −9 |
| 46 | | | 2 | | | −1.00E−09 | −9 |
| 47 | | | 2 | 2.00E−07 | | | −9 |
| 48 | | | 2 | | | | −9 |
| 49 | | 6.00E−06 | | | | | −8.6 |
| 50 | | 6.00E−06 | | −1.00E−01 | −0.8 | −2.00E−05 | 25.9 |
| 51 | | 5.00E−06 | | | −0.1 | 8.00E−06 | −5.9 |
| 52 | | 6.00E−06 | | −2.00E−01 | −0.9 | | 27.9 |
| 53 | | 5.00E−06 | | 1.00E−02 | | 6.00E−06 | −8.9 |
| 54 | | | | −9.00E−03 | −0.2 | 1.00E−04 | 6.2 |
| 55 | | 6.00E−06 | | | −0.1 | | −6.2 |
| 56 | | 6.00E−06 | | 1.00E−02 | | | −9.1 |
| 57 | | | | | −0.2 | 1.00E−04 | 4.31 |
| 58 | | 6.00E−06 | | | | −3.00E−05 | −7.8 |
| 59 | | | | 0 | −0.2 | | 5 |
| 60 | | | | 4.00E−02 | | 1.00E−04 | −5 |
| 61 | | | | | −0.2 | | 5 |
| 62 | | | | 5.00E−02 | | | −5 |
| 63 | | | | | | 0 | 1 |

## 7.5 Summary

In this chapter, we discussed the model selection procedures of discriminant analysis. Although AIC and BIC suggest the five-variable model and Cp proposes a full model by regression analysis, the "M2 minimum standard" procedure recommends a one-variable model (X1) or (X3). Revised IP-OLDF, H-SVM, SVM4, SVM1, Revised IPLP-OLDF, and logistic regression support these models. Moreover, we can explain the meaning of Theory 2 by these data. Method 2 selects X1 and X3 as two BGS naturally.

## References

Buhlmann P, Geer AB (2011) Statistics for high-dimensional data-method, theory and applications. Springer, Berlin

Cox DR (1958) The regression analysis of binary sequences (with discussion). J Roy Stat Soc B 20:215–242

# References

Firth D (1993) Bias reduction of maximum likelihood estimates. Biometrika 80:27–39

Fisher RA (1936) The use of multiple measurements in taxonomic problems. Ann Eugenics 7:179–188

Fisher RA (1956) Statistical methods and statistical inference. Hafner Publishing Co, New Zealand

Flury B, Rieduyl H (1988) Multivariate statistics: a practical approach. Cambridge University Press, Cambridge

Friedman JH (1989) Regularized discriminant analysis. J Am Stat Assoc 84(405):165–175

Goodnight JH (1978) SAS technical report—the sweep operator: its importance in statistical computing—(R100). SAS Institute Inc, USA

Jeffery IB, Higgins DG, Culhane C (2006) Comparison and evaluation of methods for generating differentially expressed gene lists from microarray data. BMC Bioinf. Jul 26 7:359:1–16. doi: 10.1186/1471-2105-7-359

Lachenbruch PA, Mickey MR (1968) Estimation of error rates in discriminant analysis. Technometrics 10:1–11

Sall JP, Creighton L, Lehman A (2004) JMP start statistics, third edition. SAS Institute Inc., USA. (Shinmura S. edits Japanese version)

Schrage L (1991) LINDO—an optimization modeling systems. The Scientific Press, UK. (Shinmura S. & Takamori, H. translate Japanese version)

Schrage L (2006) Optimization Modeling with LINGO. LINDO Systems Inc. (Shinmura S. translates Japanese version)

Shinmura S (1998) optimal linear discriminant functions using mathematical programming. J Jpn Soc Comput Stat 11/2:89–101

Shinmura S, Tarumi T (2000) Evaluation of the optimal linear discriminant functions using integer programming (IP-OLDF) for the normal random data. J Jpn Soc Comput Stat 12(2):107–123

Shinmura S (2000a) A new algorithm of the linear discriminant function using integer programming. New Trends Prob Stat 5:133–142

Shinmura S (2000b) Optimal linear discriminant function using mathematical programming. Dissertation, March 200:1–101, Okayama University, Japan

Shinmura S (2003) Enhanced algorithm of IP-OLDF. ISI2003 CD-ROM, pp 428–429

Shinmura S (2004) New algorithm of discriminant analysis using integer programming. IPSI 2004 Pescara VIP Conference CD-ROM, pp 1–18

Shinmura S (2005) New age of discriminant analysis by IP-OLDF –beyond Fisher's linear discriminant function. ISI2005, pp 1–2

Shinmura S (2007) Overviews of discriminant function by mathematical programming. J Jpn Soc Comput Stat 20(1–2):59–94

Shinmura S (2010a) The optimal linearly discriminant function. Union of Japanese Scientist and Engineer Publishing, Japan

Shinmura S (2010b) Improvement of CPU time of Revised IP-OLDF using Linear Programming. J Jpn Soc Comput Stat 22(1):39–57

Shinmura S (2011a) Beyond Fisher's linear discriminant analysi—new world of the discriminant analysis. ISI CD-ROM, pp 1–6

Shinmura S (2011b) Problems of discriminant analysis by mark sense test data. Jpn Soc Appl Stat 40(3):157–172

Shinmura S (2013) Evaluation of optimal linear discriminant function by 100-fold cross-validation. ISI CD-ROM, pp 1–6

Shinmura S (2014a) End of discriminant functions based on variance-covariance matrices. ICORE2014, pp 5–16

Shinmura S (2014b) Improvement of CPU time of linear discriminant functions based on MNM criterion by IP. Stat Optim Inf Comput 2:114–129

Shinmura S (2014c) Comparison of linear discriminant functions by $k$-fold cross-validation. Data Anal 2014:1–6

Shinmura S (2015a) The 95 % confidence intervals of error rates and discriminant coefficients. Stat Optim Inf Comput 2:66–78

Shinmura S (2015b) Four serious problems and new facts of the discriminant analysis. In: Pinson E, Valente F, Vitoriano B (ed) Operations research and enterprise systems, pp 15–30. Springer, Berlin (ISSN: 1865-0929, ISBN: 978-3-319-17508-9, doi:10.1007/978-3-319-17509-6)

Shinmura S (2015c) A trivial linear discriminant function. Stat Optim Inf Comput 3:322–335. doi:10.19139/soic.20151202

Shinmura S (2015d) Four problems of the discriminant analysis. ISI 2015:1–6

Shinmura S (2015e) The discrimination of microarray data (Ver. 1). Res Gate 1:1–4. 28 Oct 2015

Shinmura S (2015f) Feature selection of three microarray data. Res Gate 2:1–7. 1 Nov 2015

Shinmura S (2015g) Feature Selection of Microarray Data (3)—Shipp et al. Microarray Data. Research Gate (3) 1–11

Shinmura S (2015h) Validation of feature selection (4)—Alon et al. microarray data. Res Gate (4) 1–11

Shinmura S (2015i) Repeated feature selection method for microarray data (5). Res Gate 5:1–12. 9 Nov 2015

Shinmura S (2015j) Comparison Fisher's LDF by JMP and revised IP-OLDF by LINGO for microarray data (6). Res Gate 6:1–10. 11 Nov 2015

Shinmura S (2015k) Matroska trap of feature selection method (7)—Golub et al. microarray data. Res Gate (7), 18:1–14

Shinmura S (2015l) Minimum Sets of Genes of Golub et al. Microarray Data (8). Res Gate (8) 1–12. 22 Nov 2015

Shinmura S (2015m) Complete lists of small matroska in Shipp et al. microarray data (9). Res Gate (9) 1–81

Shinmura S (2015n) Sixty-nine small matroska in Golub et al. microarray data (10). Res Gate 1–58

Shinmura S (2015o) Simple structure of Alon et al. microarray data (11). Res Gate(1.1) 1–34

Shinmura S (2015p) Feature selection of Singh et al. microarray data (12). Res Gate (12) 1–89

Shinmura S (2015q) Final list of small matroska in Tian et al. microarray data. Res Gate (13) 1–160

Shinmura S (2015r) Final list of small matroska in Chiaretti et al. microarray data. Res Gate (14) 1–16

Shinmura S (2016a) The best model of swiss banknote data. Stat Optim Inf Comput, 4:118–131. International Academic Press (ISSN: 2310-5070 (online) ISSN: 2311-004X (print), doi:10.19139/soic.v4i2.178)

Shinmura S (2016b) Matroska featurE−selection method for microarray data. Biotechnology 2016:1–6

Shinmura S (2016c) The Best Model of Swiss banknote data. Statistics, Optimization and Information Computing, vol. X: 0–13

Simon N, Friedman J, Hastie T, Tibshirani R (2013) A sparsE−group lasso. J Comput Graph Stat 22:231–245

Vapnik V (1995) The nature of statistical learning theory. Springer, Berlin

# Bibliography

Alon U, Barkai N, Notterman DA, Gish K, Ybarra S, Mack D, Levine AJ, Broad AJ (1999) Patterns of gene expression revealed by clustering analysis of tumor and normal colon tissues probed by oligonucleotide arrays. Proc Natl Acad Sci USA 96:6745–6750

Chiaretti S, Li X, Gentleman R, Vitale A, Vignetti M, Mandelli F, Ritz J, Chiaretti RF (2004) Gene expression profile of adult T-cell acute lymphocytic leukemia identifies distinct subsets of patients with different response to therapy and survival. Blood, 103/7:2771–2778

Golub TR, Slonim DK, Tamayo P, Huard C, Gaasenbeek M, Mesirov JP, Coller H, Loh ML, Downing JR, Caligiuri MA, Bloomfield CD, Lander ES (1999) Molecular Classification of

Cancer: Class Discovery and Class Prediction by Gene Expression Monitoring. Science 286 (5439):531–537

Miyake A, Shinmura S (1976) Error rate of linear discriminant function. In: Dombal FT, Gremy F (ed). North-Holland Publishing Company, The Netherland, pp 435–445

Sall JP (1981) SAS regression applications. SAS Institute Inc., USA. (Shinmura S. translate Japanese version)

Shipp MA, Ross KN, Tamayo P, Weng AP, Kutok JL, Aguiar RC, Gaasenbeek M, Angelo M, Reich M, Pinkus GS, Ray TS, Koval MA, Last KW, Norton A, Lister TA, Mesirov J, Neuberg DS, Lander ES, Aster JC, Golub TR (2002) Diffuse large B-cell lymphoma outcome prediction by genE–expression profiling and supervised machine learning. Nat Med 8:68–74

Singh D, Febbo PG, Ross K, Jackson DG, Manola J, Ladd C, Tamayo P, Renshaw AA, D'Amico A, Richie JP, Lander ES, Lada M, Kantoff PW, Golub TR, Sellers WR (2002) Gene expression correlates of clinical prostate cancer behavior. Cancer Cell 1:203–209

Taguchi G, Jugular R (2002) The mahalanobis-taguchi strategy—a pattern technology system. Wiley, New York

Tian E, Zhan F, Walker R, Rasmussen E, Ma Y, Barlogie B, Shaughnessy JD (2003) The role of the wnt-signaling antagonist dkk1 in the development of osteolytic lesions in multiple myeloma. New Engl J Med 349(26):2483–2494

# Chapter 8
# Matroska Feature-Selection Method for Microarray Dataset (Method 2)

## 8.1 Introduction

In this chapter, we introduce the Matroska feature-selection method for microarray dataset (Method 2). We have already established a new theory of discriminant analysis (Theory) and developed five optimal linear discriminant functions (OLDFs) in Chap. 1. First, we developed OLDF using integer programming (IP-OLDF) based on minimum number of misclassifications (MNM) criterion by IP. It reveals two new facts about discriminant analysis. Because IP-OLDF has a defect for Problem 1, Revised IP-OLDF is proposed (Shinmura 2007, 2010a, 2011a, b, 2013, 2014c).

Discriminant analysis has five serious problems (Shinmura 2014a, 2015c, d). We could not discriminate cases on the discriminant hyperplane (Problem 1). Only Revised IP-OLDF could solve this problem theoretically. Only a hard-margin SVM (H-SVM) (Vapnik 1995) and Revised IP-OLDF could discriminate the linearly separable dataset (LSD) theoretically (Problem 2).[1] Problem 3 was that the generalized inverse matrices technique of variance–covariance matrices and QDF misclassified all cases to another class for a particular case. We solved Problem 3 by adding small random noises for a constant value. Because Fisher never formulated the standard-error (SE) equation of the discriminant coefficient and error rate (Miyake and Shinmura 1976), discriminant analysis is not the traditional inferential statistical theory (Problem 4). We developed the 100-fold cross-validation for a small sample method (Method 1) instead of the leave-one-out (LOO) procedure (Lachenbruch and Mickey 1968). The Method 1 offers a 95 % CI for the error rate and coefficient (Shinmura 2015a, b). Moreover, we obtained two means of error rates, M1 and $M2$, in the training and validation samples and proposed a simple model selection procedure to choose the best model with a minimum $M2$. We compared two statistical LDFs and six MP-based LDFs: Fisher's LDF (Fisher 1936,

---

[1]We determined experimentally that Revised LP-OLD could discriminate LSD correctly.

1956), logistic regression (Cox 1958, Firth 1993), H-SVM, two S-SVMs, Revised IP-OLDF, and another two OLDFs. The best model of Revised IP-OLDF was found to be better than the other seven best models (*M*2s) in the six different types of data. Because we solved four problems completely in 2015 (Shinmura 2016a, c), we misunderstood to establish Theory in 2015.

At the end of October 2015, we discriminated six microarray datasets (the datasets) using six MP-based LDFs and Fisher's LDF. The JMP statistical package (Sall et al. 2004) currently does not support logistic regression for the dataset. Because the NMs of Fisher's LDF are not zero, we cannot use Fisher's LDF to analyze the dataset (Shinmura 2015e–s, 2016b). Although the NMs of three SVMs are zero, most coefficients of three SVMs are not zero. Therefore, three SVMs are not helpful for feature-selection. Because several coefficients of three OLDFs are not zero, and most of the coefficients are zero, we find three OLDFs can select the dataset features within a few seconds and reduce high-dimensional gene spaces to smaller subspaces. Because Revised LP-OLDF and Revised IPLP-OLDF cannot discriminate LSD theoretically and we would like to finalize the Theory 2, we no longer consider the two OLDFs for feature-selection.

In this chapter, we call the linearly separable gene space or subspaces "Matroskas," referring to the Russian dolls that nest inside each other. The dataset is the big Matroska, which includes a huge number of smaller Matroskas. Only Revised IP-OLDF can explain the structure of the dataset using the Method 2, as follows. Because several coefficients of Revised IP-OLDF are not zero, and most of the coefficients are zero, we find Revised IP-OLDF can select the dataset features within a few seconds and reduce the high-dimensional gene space (big Matroska) to smaller subspaces (Matroskas).

Next, when we discriminate the subspace again, we can find the smaller subspaces. Therefore, we find the dataset has the structure of a Matroska. When we cannot find any smaller Matroskas, we call the last subspace the small Matroska (SM). Moreover, after we exclude the first SM from the dataset, we find the second SM. Finally, we can list all SMs and conclude that the dataset consists of several disjoint SMs and another high-dimensional gene subspace that is not linearly separable.

We confirmed that Revised IP-OLDF could naturally select features for Swiss banknote data (Flury and Rieduyl 1988) in Chap. 6 and Japanese-automobile data (Shinmura 2016c) in Chap. 7, in addition to the six datasets. We developed the Method 2 using the LINGO program (Schrage 2006). Therefore, we can list all SMs for the six datasets (Alon et al. 1999; Chiaretti et al. 2004; Golub et al. 1999; Shipp et al. 2002; Singh et al. 2002; Tian et al. 2003).

For more than ten years, researchers have struggled to analyze the dataset because the genes are huge (Problem 5). However, we find the dataset consists of several disjoint SMs; their MNMs are zero. We can analyze these SMs very easily because each SM is a small sample. We have established our Theory, and it is the most helpful for feature-selection.

## 8.1 Introduction

Recently, many researchers have approached this theme using LASSO or ordinary statistical methods (Buhlmann and Geer 2011; Jeffery et al. 2006; Simon et al. 2013). In addition to Chaps. 6 and 7, this chapter offer useful datasets and results for these researchers from the following points:

1. Because Revised IP-OLDF can naturally select features for Swiss banknote and Japanese-automobile data, they had better discriminate these data by their LDFs. If their LDFs are not successful for ordinary data, I think those are not successful for the dataset. Because I had already showed Fisher's LDF could not discriminate LSD correctly, it cannot discriminate the dataset correctly. However, it is very helpful for us that JMP develops Fisher's LDF for the dataset because I can confirm my claim and save research time. On the other hand, researchers should be as soon as possible determining the likelihood of their own research theme.
2. Can their LDFs discriminate our eight different datasets exactly?
3. Can their LDFs find the Matroska structure correctly and list all basic gene sets or subspaces (BGSs), including the smallest Matroska in each SM?

## 8.2 Matroska Feature-Selection Method (Method 2)

### 8.2.1 Short Story to Establish Method 2

We regret that we did not discriminate the dataset more earlier. We developed the Theory of discriminant analysis using IP from 1997 to 2010 (Shinmura 2010a). Many statistical researchers cannot understand our research because they assumed a normal distribution. In the MP society, most researchers are not interested in statistics; nevertheless, there were many MP-based discriminant functions before Stam (1997). However, both researchers know about SVM, although H-SVM and S-SVM are LDFs using QP (Vapnik 1995).

When we discriminated Shipp et al.'s dataset (2002) on Oct. 28, 2015, only a few coefficients of Revised IP-OLDF were not zero, and many other coefficients became zero (Shinmura 2015e). We had already found that Revised IP-OLDF could naturally select features for the Swiss banknote data in Chap. 6 and the Japanese-car data in Chap. 7. Only Revised IP-OLDF using IP could select a few features from among 7129 genes. Although we misunderstood that discrimination with 7129 variables was difficult before 2010, Fisher's LDF using JMP ver. 12 (JMP12) and other MP-based LDFs, coded using LINGO as described in Sect. 2.3.3, can solve the datasets in less than 20 s because the datasets are LSD.

However, most coefficients of three SVMs, except for Revised IP-OLDF, are not zero. Therefore, these SVMs are not helpful for gene-feature-selection. Moreover, we found that gene spaces are the disjoint union of SMs with "MNM = 0" and another high-dimensional gene subspace with "MNM $\geq$ 1." If we can list all of the disjoint SMs, we can completely understand the dataset structure. We can analyze each SM using ordinary statistical methods because the SM case numbers are generally less than sample case number and the number of genes of all SMs are less than 104 for the six datasets. These datasets are small samples in statistics. Many researchers can easily analyze these SMs using ordinary statistical methods such as *t* test, one-way ANOVA, clustering, and PCA.

Because of our breakthrough, the dataset feature-selection is a very easy and exciting theme. We are grateful that Jeffery et al. (2006) uploaded six datasets on HP.[2] Their paper tells us about the outlook of feature-selection methods until now. Moreover, we confirmed that our Theory is superior to ordinary statistical approaches and LASSO.

First, although three OLDFs, three SVMs, and Fisher's LDF discriminated the Shipp et al. dataset, we only discuss Revised IP-OLDF, H-SVM, and Fisher's LDF in this chapter. Only Revised IP-OLDF can naturally select features. We present two reasons why Revised IP-OLDF may have this surprising superiority to other LDFs.

1. The MNM criterion works well for feature-selection. Moreover, Revised LP-OLDF and Revised IPLP-OLDF show feature-selection. However, three SVMs cannot make feature-selection. Therefore, we discuss on, H-SVM, and Fisher's LDF in this chapter.
2. The LINGO IP-solve algorithm uses branch and bound. We believe that Revised IP-OLDF coded by another IP algorithm cannot select features naturally. However, we cannot control the flow of the branch and bound. When the IP solver first finds the model with MNM = 0, LINGO outputs it and ends by the "first-in, first-out rule." Therefore, we cannot find the smallest basic gene subspaces (BGSs) in the SM. If we know the BGSs, we can completely understand the dataset structure using BGS because of the monotonic decreasing of MNM.

Next, we discriminate two datasets, Alon et al. and Golub et al., using the above LDFs and confirm the feature-selection by Revised IP-OLDF (Shinmura 2015f).[3] In the third step, we propose the first version of our feature-selection method (Shinmura 2015g, h). However, we know the omitted genes are linearly separable models and find the disjoint structure of the Matroska (Shinmura 2015i) in the fourth step. On Nov. 11, 2015, JMP released JMP ver. 12 (JMP12), which can discriminate dataset using Fisher's LDF; we borrowed it from the JMP division of

---

[2]http://www.bioinf.ucd.ie/people/ian/.

[3]In our research, IP-OLDF (Shinmura 1998, 2000a, b, 2003, 2004, 2005; Shinmura and Tarumi 2000) was proposed before Revised IP-OLDF. It reveals the following two important facts about discriminant analysis:

1. The optimal convex polyhedron (optimal CP, OCP) with MNM;
2. MNM decreases monotonically ($MNM_k \geq MNM_{(k+1)}$). If $MNM_k = 0$, numerous models, including these k-variables, are linearly separable models in 7,129-dimensional space. Namely, although there are numerous linearly separable models in the gene space, we can understand the dataset structure using these SMs. However, because BGSs can list all linearly separable models (Matroska) completely, our target is to list all the BGSs in SMs.

## 8.2 Matroska Feature-Selection Method (Method 2)

**Table 8.1** Summary of six microarray dataset (Nov. 2015)

| Dataset | Alon et al. (1999) | Chiaretti et al. (2004) | Golub et al. (1999) | Shipp et al. (2002) | Singh et al. (2002) | Tian et al. (2003) |
|---|---|---|---|---|---|---|
| SM: Gene | 64:1152 | 270:5385 | 69:1238 | 213:3032 | 179:3990 | 159:7221 |
| Min, Mean, Max | 11 18 39 | 9 19 62 | 10 18 31 | 7 14 43 | 13 22 47 | 28 45.4 104 |
| JMP12 | 20:2/3:37 | 94:1/2:31 | 20:5/3:44 | 17:2/1:57 | 46:4/6:46 | 16:20/9:128 |
| % and error rate | 63, 8 % | 49, 1 % | 43, 11 % | 56, 4 % | 46, 10 % | 60, 17 % |

SAS Institute Japan. However, the NMs of Fisher's LDF are not zero in Table 8.1 (Shinmura 2015j). In the fifth step, we propose the second version of the Matroska feature-selection method using manual operation (Shinmura 2015k–l). Because we fear that there are some mistakes in the manual operation, we developed a third version of Method 2, programmed using LINGO (Shinmura 2015m–r) that is LINGO Program 3. Six papers (Shinmura 2015m–r) are helpful for gene analysis because those papers include full lists of SMs.

### 8.2.2 Explanation of Method 2 by Alon et al. Dataset

#### 8.2.2.1 Feature-Selection by Eight LDFs

Many researchers are struggling with dataset feature-selection. Jeffery et al. (2006) compared the efficiency of ten feature-selection methods using ordinary statistical approaches. They uploaded six different two-class datasets. We discriminated those datasets using six MP-based LDFs. Table 8.2 shows the results. Although Chiaretti et al. dataset includes four two-class discrimination problems, we use only "B cell or T cell" as the two-class object variable. All CPU times are less than 36 s. H-SVM, SVM4 (penalty $c = 10{,}000$), and SVM1 (penalty $c = 1$) cannot select features for the five datasets except for the Golub et al. dataset. Only 903 and 904 coefficients of H-SVM, and SVM1, respectively, become zero for the Golub et al. dataset. Because all NMs of Fisher's LDF are not zero, it is not useful for gene analysis.

JMP12 does not support logistic regression for high-dimensional dataset. Because logistic regression can often discriminate LSD, it may be able to select dataset features. Although three SVMs can recognize these datasets as LSD, the three SVMs are not helpful for feature-selection of these datasets. Maximizing the distance between two support vectors (SVs) may cause these results, because three OLDFs can naturally select features for the ordinary data in Chaps. 6 and 7. Although Revised IPLP-OLDF (IPLP) and Revised LP-OLDF (LP) can select features better than Revised IP-OLDF (RIP) because their nonzero coefficients are fewer than RIP in the first discrimination, these two OLDFs cannot find smaller Matroskas after the second discrimination. This is my future work.

**Table 8.2** Summary of six datasets (*left* CPU time; *right* number of nonzero coefficients)

|      | Alon et al. |      | Chiaretti et al. |        | Golub et al. |      | Shipp et al. |      | Singh et al. |        | Tian et al. |        |
|------|-----|------|-----|--------|-----|------|-----|------|------|--------|------|--------|
| RIP  | 0s  | 62   | 11s | 128    | 1s  | 72   | 4s  | 65   | 36s  | 92     | 14s  | 173    |
| IPLP | 1s  | 40   | 6s  | 38     | 2s  | 27   | 2s  | 22   | 4s   | 75     | 9s   | 118    |
| LP   | 1s  | 40   | 4s  | 38     | 1s  | 622  | 2s  | 22   | 2s   | 75     | 5s   | 139    |
| HSVM | 0s  | 2000 | 4s  | 12,625 | 1s  | 6226 | 3s  | 7129 | 2s   | 12,625 | 8s   | 12,625 |
| SVM4 | 0s  | 2000 | 6s  | 12,625 | 1s  | 7129 | 3s  | 7129 | 4s   | 12,625 | 8s   | 12,625 |
| SVM1 | 0s  | 2000 | 7s  | 12,625 | 2s  | 6225 | 3s  | 7129 | 2s   | 12,625 | 7s   | 12,625 |

Many statisticians struggled with analyzing high-dimensional gene datasets using ordinary statistical approaches and now expect LASSO (Simon et al. 2013) to select features. However, because MP-based LDFs can already select features easily, we suggest that they compare their methods with our results in Table 8.2.

### 8.2.2.2 Results of Alon et al. Dataset Using the LINGO Program

Revised LP-OLDF and Revised IPLP-OLDF can easily reduce 2000 gene spaces to forty-gene subspaces; however, three SVMs cannot naturally select features. When both OLDFs discriminate the forty genes again, they cannot reduce the forty-gene subspace to a smaller gene subspace. On the other hand, Revised IP-OLDF can reduce a 62-gene subspace to a 43-gene subspace and reduce a 43-gene subspace to a 29-gene subspace. Therefore, the Alon et al. dataset has a Matroska series: Alon2000 ⊃ Alon62 ⊃ Alon43 ⊃ Alon29. Therefore, only Revised IP-OLDF can achieve the Theory 2. We call Alon2000 the big Matroska and Alon29 a small Matroska (SM). If we can find a smallest Matroska (BGS) in SM, we can understand the structure of the Alon et al. dataset by the monotonic decrease of the MNM.

For this section, we discriminated six datasets with the "LINGO Program 3 of Method 2" using a Dell Optiplex 3020 (Windows 10) in 2016. The results after Sect. 8.3 were obtained by a Dell Vostro (Windows 7) in 2015. Although the branch-and-bound algorithm of IP solver may output different optimal solutions (e.g., the first-in first-out rule), it is important that the BGSs are the same in each dataset.

Table 8.3 shows the first and second SMs among 66 SMs; nevertheless, Table 8.1 shows that Vostro computed 64 SMs in 2015.[4] The first SM1 and second SM2 consist of 15 genes. In addition to the different PCs, we changed the number of reducing iterations from 11 to five in this section to reduce the CPU times. The "LINGO Program 3" finds these 15 genes in SM1 and excludes them from the 2000

---

[4]By manual operation, LINGO finds Alon29 as the first SM. However, "LINGO Program 3" finds the first SM1 with 15 genes.

## 8.2 Matroska Feature-Selection Method (Method 2)

**Table 8.3** Two gene lists of first and second SMs

| SN | Loop1 |||| 
|---|---|---|---|---|
| | 1 || 2 ||
| | Var. | Gene | Var. | Gene |
| 1 | X6 | Hsa.20,836 | X2 | Hsa.13491 |
| 2 | X11 | Hsa.750 | X59 | Hsa.1732 |
| 3 | X23 | Hsa.3002 | X175 | Hsa.18664 |
| 4 | X24 | Hsa.1119 | X365 | Hsa.821 |
| 5 | X178 | Hsa.10510 | X556 | Hsa.268 |
| 6 | X251 | Hsa.41315 | X660 | Hsa.347 |
| 7 | X713 | Hsa.15844 | X1258 | Hsa.5211 |
| 8 | X969 | Hsa.35804 | X1346 | Hsa.5392 |
| 9 | X976 | Hsa.25867 | X1370 | Hsa.35496 |
| 10 | X1367 | Hsa.7648 | X1473 | Hsa.1410 |
| 11 | X1671 | Hsa.627 | X1505 | Hsa.862 |
| 12 | X1799 | Hsa.23824 | X1521 | Hsa.295 |
| 13 | X1812 | Hsa.41369 | X1733 | Hsa.2749 |
| 14 | X1966 | Hsa.36705 | X1804 | Hsa.2199 |
| 15 | X1986 | Hsa.2484 | X1932 | Hsa.2243 |

genes. In the second big iteration (Loop1), Revised IP-OLDF discriminates a second big Matroska with 1985 genes and outputs the second SM2 with Loop1 = 2 (SM2). In this book, the original gene name is replaced by the simple variable $X$ with a suffix. $X1$ to $X2000$ correspond to genes, and $X2001$ or "$c$" refers to the intercept term of Revised IP-OLDF. This rule is the same for the six datasets. If a researcher wishes to analyze one of the SMs, she/he should add the simple variable $X$ with a suffix to the downloaded datasets.

Table 8.4 shows a list of 66 SMs with "MNM = 0" and other gene subspaces from SM67 to SM82. Column "$n$" means the number of genes and "Cum." means the cumulative number of genes. After SM66, there are 16 gene subspaces with "MNM ≥ 1." We consider that MNM shows the priorities of cancer diagnosis. We need not focus on the 16 sms prior to 66 SMs. Because the six datasets are LSD, we focus on the SMs with "MNM = 0." However, MNM also tells us the priority for subspaces that are not linearly separable. Therefore, the Method 2 is useful for other gene datasets that are not LSD.

### 8.2.3 Summary of Six Microarray Datasets in 2016

We downloaded six datasets from Jeffery et al. (2006). Their paper tells us about the outlook of feature-selection methods. We developed the "LINGO Program 3 of Method 2" to find all SMs amd sms. We are free from the mistakes of using manual operation. We had already analyzed the six datasets in Table 8.1. Table 8.5 summarizes the six datasets computed in May 2016 under a different PC environment.

**Table 8.4** List of 66 SMs and other subspaces

| SM | MNM | $n$ | Cum. |
| --- | --- | --- | --- |
| SM1 | 0 | 15 | 15 |
| SM2 | 0 | 15 | 30 |
| SM3 | 0 | 11 | 41 |
| SM4 | 0 | 11 | 52 |
| SM5 | 0 | 14 | 66 |
| SM6 | 0 | 19 | 85 |
| SM7 | 0 | 11 | 96 |
| SM8 | 0 | 14 | 110 |
| SM9 | 0 | 12 | 122 |
| SM10 | 0 | 18 | 140 |
| SM11 | 0 | 14 | 154 |
| SM12 | 0 | 16 | 170 |
| SM13 | 0 | 14 | 184 |
| SM14 | 0 | 16 | 200 |
| SM15 | 0 | 14 | 214 |
| SM16 | 0 | 14 | 228 |
| SM17 | 0 | 18 | 246 |
| SM18 | 0 | 14 | 260 |
| SM19 | 0 | 14 | 274 |
| SM20 | 0 | 20 | 294 |
| SM21 | 0 | 14 | 308 |
| SM22 | 0 | 12 | 320 |
| SM23 | 0 | 15 | 335 |
| SM24 | 0 | 14 | 349 |
| SM25 | 0 | 14 | 363 |
| SM26 | 0 | 16 | 379 |
| SM27 | 0 | 17 | 396 |
| SM28 | 0 | 16 | 412 |
| SM29 | 0 | 16 | 428 |
| SM30 | 0 | 14 | 442 |
| SM31 | 0 | 16 | 458 |
| SM32 | 0 | 15 | 473 |
| SM33 | 0 | 17 | 490 |
| SM34 | 0 | 17 | 507 |
| SM35 | 0 | 14 | 521 |
| SM36 | 0 | 17 | 538 |
| SM37 | 0 | 19 | 557 |
| SM38 | 0 | 15 | 572 |
| SM39 | 0 | 16 | 588 |
| SM40 | 0 | 16 | 604 |
| SM41 | 0 | 17 | 621 |

(continued)

## 8.2 Matroska Feature-Selection Method (Method 2)

**Table 8.4** (continued)

| SM | MNM | $n$ | Cum. |
|---|---|---|---|
| SM42 | 0 | 16 | 637 |
| SM43 | 0 | 17 | 654 |
| SM44 | 0 | 19 | 673 |
| SM45 | 0 | 16 | 689 |
| SM46 | 0 | 13 | 702 |
| SM47 | 0 | 15 | 717 |
| SM48 | 0 | 17 | 734 |
| SM49 | 0 | 18 | 752 |
| SM50 | 0 | 22 | 774 |
| SM51 | 0 | 18 | 792 |
| SM52 | 0 | 22 | 814 |
| SM53 | 0 | 20 | 834 |
| SM54 | 0 | 15 | 849 |
| SM55 | 0 | 19 | 868 |
| SM56 | 0 | 21 | 889 |
| SM57 | 0 | 23 | 912 |
| SM58 | 0 | 19 | 931 |
| SM59 | 0 | 20 | 951 |
| SM60 | 0 | 21 | 972 |
| SM61 | 0 | 22 | 994 |
| SM62 | 0 | 26 | 1020 |
| SM63 | 0 | 22 | 1042 |
| SM64 | 0 | 29 | 1071 |
| SM65 | 0 | 28 | 1099 |
| SM66 | 0 | 32 | 1131 |
| sm67 | 1 | 19 | 1150 |
| sm68 | 1 | 21 | 1171 |
| sm69 | 1 | 21 | 1192 |
| sm70 | 1 | 29 | 1221 |
| sm71 | 1 | 22 | 1243 |
| sm72 | 1 | 26 | 1269 |
| sm73 | 1 | 18 | 1287 |
| sm74 | 1 | 27 | 1314 |
| sm75 | 1 | 34 | 1348 |
| sm76 | 2 | 31 | 1379 |
| sm77 | 2 | 25 | 1404 |
| sm78 | 2 | 30 | 1434 |
| sm79 | 3 | 26 | 1460 |
| sm80 | 3 | 18 | 1478 |
| sm81 | 4 | 31 | 1509 |
| sm82 | >5 | 401 | 2000 |

Table 8.5 Summary of six microarray datasets (May 2016)

| Dataset | Alon et al. (1999) | Chiaretti et al. (2004) | Golub et al. (1999) | Shipp et al. (2002) | Singh et al. (2002) | Tian et al. (2003) |
|---|---|---|---|---|---|---|
| Description | Normal (22) versus tumor cancer (40) | B cell (95) versus T cell (33) | All (47) versus AML (25) | Follicular lymphoma (19) versus DLBCL (58) | Normal (50) versus tumor prostate (50) | False (36) versus true (137) |
| Size | 62 × 2000 | 128 × 12625 | 72 × 7129 | 77 × 7129 | 102 × 12625 | 173 × 12625 |
| SM: Gene | 66:1131 | 269:5220 | 67:1203 | 214:3040 | 178:3984 | 159:7221 |
| Min, Mean, Max | 11, 17.1, 32 | 9, 19.4, 71 | 10, 19.4, 41 | 7, 14.2, 39 | 13, 22.4, 46 | 28, 48, 152 |

The "Description" rows show two classes. The "Size" rows are the case number by the gene number. The "SM: Gene" rows are the number of SMs, and the number of genes included in all SMs. The "Min, Mean, Max" row is the minimum, mean, and maximum values of the genes included in all SMs.

### 8.2.4 Summary of Six Datasets in 2015

The "LINGO Program 3" had already analyzed the six datasets in November 2015. Table 8.1 shows the summary. The "SM: Gene" rows are the number of SMs and the number of genes included in all SMs. Six references (Shinmura 2015m–r) include the full gene name of all SMs in the six datasets and are the most important among the 15 papers. Researchers can find the BGSs in each SM using a manual survey or by analyzing SMs with ordinary statistical methods. The "Min, Mean, Max" rows are the minimum, mean, and maximum values of the genes included in all SMs.

The numbers in Tables 8.1 and 8.5 are slightly different because the computational environments are different. However, each dataset has the same BGSs. Because we had already uploaded full lists of all SMs of the six dataset, we explain the Theory 2 detail using the results in 2015 after this section. The "JMP12" rows are 2 × 2 tables of the discrimination by Fisher's LDF for high-dimensional dataset (Shinmura 2015j). Because the six NMs are 5, 3, 8, 3, 10, and 29, the error rates are high. The "% and error rate" rows are the percentages of (maximum value/case number) and the error rates of JMP12. The maximum percentage is 63 % for the Alon et al. dataset. The minimum percentage is 43 % by the Golub et al. dataset. The maximum error rate is 17 % by the Tian et al. dataset, and the minimum error rate is 1 % by the Chiaretti et al. dataset. In the future, we expect these numbers might show the characteristics of the dataset.

We must be aware of the optimal solutions from MP-based LDFs that output only one optimal solution, even though there are several optimal solutions. However, we can obtain the $k$-best solutions using the "$k$-best option in Sect. 4.2.4 of the IP solver." However, this verification is outside the scope of this book.

## 8.3 Results of the Golub et al. Dataset

### 8.3.1 Outlook of Method 2 by the LINGO Program 3

We introduce the outlook of the Method 2 again. Table 8.6 shows the output of the Golub et al. dataset. The "Loop1 and Loop2" columns are the sequence numbers for the big and small loops of the Method 2, respectively. Revised IP-OLDF discriminates the dataset with 7129 genes in Loop1 = 1 and Loop2 = 1, and only 34 coefficients of Revised IP-OLDF are not zero. In general, this number is less than

**Table 8.6** Outlook of theory 2

| SN | Loop1 | Loop2 | Gene | MNM |
|---|---|---|---|---|
| 1 | 1 | 1 | 7129 | 0 |
| 2 | 1 | 2 | 34 | 0 |
| 3 | 1 | 3 | 11 | 0 |
| 4 | 1 | 4 | 11 | 0 |
| 16 | 2 | 1 | 7118 | 0 |
| 17 | 2 | 2 | 36 | 0 |
| 18 | 2 | 3 | 18 | 0 |
| 19 | 2 | 4 | 16 | 0 |
| 20 | 2 | 5 | 16 | 0 |

the case number, e.g., 72. In the second small loop (Loop1 = 1, Loop2 = 2), we discriminate the smaller Matroska with 34 genes, and only 11 coefficients are not zero. Therefore, we get a Matroska series (or Matroska product), e.g., Golub7129 → Golub34 → Golub11. We stop at Loop2 = 4 because we cannot find the smaller Matroska in the fourth small loop. We call Golub11 the first SM1 because Revised IP-OLDF cannot locate a smaller Matroska. By the definition of the Matroska series, we can produce the first Matroska, product1, if we are a Matroska producer. We exclude SM1 with 11 genes from the big Matroska with 7129 genes and make the second big Matroska with 7118 genes. In the second big loop at Loop1 = 2, we find the second SM2 with 16 genes at Loop2 = 5. We stop the big loop when we find an MNM that is greater than one at Loop1 = 70. Therefore, we find 69 SMs from SM1 to SM69 in the Golub et al. dataset. However, we can continue these big and small loops after finding SM69. In this table, because the maximum number of small iterations is fixed to 15, SN starts with 16 in Loop1 = 2. Because the different maximum numbers may find the different lists of all SMs in the dataset, I must develop a "Revised LINGO Program 3" to find all BGSs in the dataset as soon as possible. Different from the SMs, BGSs have a unique list.

Table 8.7 is the list of all SMs. The "Loop1" column is the sequential number of 69 SMs from SM1 (Loop1 = 1) to SM69 (Loop1 = 69). Therefore, the dataset consists of disjoint unions of 69 SMs, and all MNMs are zero. The "Gene" column is the number of genes in the big Matroska at Loop2 = 1. First, Revised IP-OLDF discriminates Golub7129 and finds Golub11 as SM1. Next, Revised IP-OLDF discriminates Golub7118 without SM1 and finds Golub16 as SM2. The "N_SM" column is the number of genes in each SM. The "logistic" column shows the linearly separable subspace found by logistic regression.[5] The "N_BGS" column shows the gene number of BGS. Therefore, SM1 includes one BGS with four

---
[5]Logistic regression becomes unstable for the discrimination of LSD and SE of logistic coefficients become large values (Firth 1993). If it satisfies two conditions, I judge it can discriminate LSD correctly. Two conditions are as follows: (1) NM of it is zero if I search the minimum NM on ROC, and (2) MNM = 0. It can almost discriminate LSD and is better than Fisher's LDF, QDF, and RDA. However, it cannot select features for ordinal data in this book.

## 8.3 Results of the Golub et al. Dataset

**Table 8.7** 69 SMs of the Golub et al. Dataset

| Loop1 | Gene | N_SM | Logistic | N_BGS |
|---|---|---|---|---|
| 1 | 7129 | 11 | 4 | 4 |
| 2 | 7118 | 16 | 8 | 6 |
| 3 | 7102 | 11 | 6 | 6 |
| 4 | 7091 | 10 | 4 | 3 |
| 5 | 7081 | 13 | 7 | 5 |
| 6 | 7068 | 12 | 9 | 9 |
| 7 | 7056 | 13 | 8 | |
| 8 | 7043 | 12 | 5 | 4 |
| 9 | 7031 | 14 | 5 | 5 |
| 10 | 7017 | 16 | 7 | 6 |
| 11 | 7001 | 10 | 7 | 6 |
| 12 | 6991 | 12 | 8 | 8, 2 |
| 13 | 6979 | 13 | 7 | 5 |
| 14 | 6966 | 16 | 15 | |
| 15 | 6950 | 14 | 7 | 7 |
| 16 | 6936 | 13 | 8 | |
| 17 | 6923 | 19 | 10 | 9 |
| 18 | 6904 | 15 | 8 | 8 |
| 19 | 6889 | 13 | 9 | 9 |
| 20 | 6876 | 14 | 8 | 7 |
| 21 | 6862 | 16 | 14 | |
| 22 | 6846 | 17 | 9 | 9 |
| 23 | 6829 | 17 | 14 | |
| 24 | 6812 | 14 | 10 | 10 |
| 25 | 6798 | 16 | 12 | 10 |
| 26 | 6782 | 15 | 12 | 12, 3 |
| 27 | 6767 | 12 | 7 | 6 |
| 28 | 6755 | 21 | 20 | |
| 29 | 6734 | 15 | 10 | 9 |
| 30 | 6719 | 14 | 10 | 9 |
| 31 | 6705 | 22 | 11 | 10 |
| 32 | 6683 | 19 | 14 | 13 |
| 33 | 6664 | 16 | 13 | |
| 34 | 6648 | 18 | 14 | |
| 35 | 6630 | 17 | 14 | |
| 36 | 6613 | 19 | 17 | |
| 37 | 6594 | 12 | 9 | |
| 38 | 6582 | 16 | 12 | 12 |
| 39 | 6566 | 16 | 13 | |
| 40 | 6550 | 16 | 15 | |
| 41 | 6534 | 19 | 12 | 12 |

(continued)

**Table 8.7** (continued)

| Loop1 | Gene | N_SM | Logistic | N_BGS |
|---|---|---|---|---|
| 42 | 6515 | 14 | 14 | 11 |
| 43 | 6501 | 19 | 19 | |
| 44 | 6482 | 14 | 14 | |
| 45 | 6468 | 21 | 18 | |
| 46 | 6447 | 21 | 18 | |
| 47 | 6426 | 20 | 20 | |
| 48 | 6406 | 23 | 20 | |
| 49 | 6383 | 19 | 17 | |
| 50 | 6364 | 19 | 16 | |
| 51 | 6345 | 24 | 15 | |
| 52 | 6321 | 19 | 18 | |
| 53 | 6302 | 20 | 15 | |
| 54 | 6282 | 22 | 22 | |
| 55 | 6260 | 19 | 18 | |
| 56 | 6241 | 24 | 23 | |
| 57 | 6217 | 21 | 18 | |
| 58 | 6196 | 25 | 20 | |
| 59 | 6171 | 27 | 26 | |
| 60 | 6144 | 20 | 20 | |
| 61 | 6124 | 23 | 22 | |
| 62 | 6101 | 28 | 25 | |
| 63 | 6073 | 23 | 20 | |
| 64 | 6050 | 23 | 23 | |
| 65 | 6027 | 28 | 28 | 28 |
| 66 | 5999 | 23 | 23 | |
| 67 | 5976 | 23 | 23 | |
| 68 | 5953 | 31 | 26 | 25 |
| 69 | 5922 | 31 | 31 | 30 |
| Other | 5891 | | | |

genes, and SM2 includes one BGS with six genes. SM12 has two BGSs with eight and two genes.

Although most researchers struggle with high-dimensional gene spaces using statistical methods or LASSO, it is very easy for us to analyze each SM because SM68 and SM69 are the biggest samples, with 72 cases by 31 genes among 69 SMs. Although we can find BGS in 32 SMs by manual operation, we have stopped this work. In the near future, we will develop "Revised LINGO Program 3."

8.3 Results of the Golub et al. Dataset

Our primary concern is to find the disjoint union of SMs among 7129 genes. We omit the high-dimensional subspace that is not linearly separable (MNM ≥ 1). SM1 contains eleven genes. SM69 contains 31 genes. 69 SMs contain 1238 genes. We omit the detail of another 5891 genes that are not linearly separable.

### 8.3.2 First Trial to Find the Basic Gene Sets

Because we cannot control the flow of the branch-and-bound algorithm of the LINGO IP solver, there may be smallest linearly separable models or subspaces in the SM. We call these smallest Matroska subspaces in each SM "BGSs." We propose finding BGSs as follows:

1. We analyze SM1 with 11 genes by the forward-stepwise procedure and obtain the six columns from "Step" to "BIC" of Table 8.8. The last column is the NM of logistic regression because it looks for the linearly separable model better than Fisher's LDF, QDF, and RDA (Friedman 1989) by JMP. We know the four-variable model is linearly separable. Mallow's $C_p$, Akaike's information criterion (AIC), and the Bayesian information criteria (BIC) with bold statistics recommend this model among eleven models. Usually, these three statistics recommend different models. Therefore, I think two classes of the dataset are more separated than ordinary data discriminated by me until now.
2. We search the BGSs using all possible combinations of four variables (Goodnight 1978) using Revised IP-OLDF in Table 8.9.

Table 8.9 includes 15 models by four genes from the second column (Y1) to the fifth column (Y4). The "c" column is the intercept of Revised IP-OLDF. The "p" column is the number of independent variables from four variables (p = 4) to four one-variable models (p = 1). Binary values such as 1/0 indicate whether or not each model includes each variable in the model. The column "MNM" is the MNM of the

**Table 8.8** Forward stepwise and logistic regression

| Step | Gene | Var. | Cp | AIC | BIC | Logistic |
|---|---|---|---|---|---|---|
| 1 | M11722_at | Y1 | 72.56 | 137.78 | 144.26 | 5 |
| 2 | X59871_at | Y2 | 38.42 | 118.62 | 127.13 | 2 |
| 3 | U05259_rna1_at | Y3 | 9.92 | 96.07 | 106.54 | 2 |
| 4 | D21063_at | Y4 | **3.88** | **90.15** | **102.52** | 0 |
| 5 | M22919_rna2_at | | 3.80 | 90.30 | 104.49 | 0 |
| 6 | M21624_at | | 4.27 | 91.09 | 107.02 | 0 |
| 7 | M25280_at | | 4.63 | 91.79 | 109.38 | 0 |
| 8 | L13210_at | | 6.15 | 93.93 | 113.09 | 0 |
| 9 | X82240_rna1_at | | 8.02 | 96.56 | 117.21 | 0 |
| 10 | HG3039-HT3200_at | | 10.01 | 99.44 | 121.47 | 0 |
| 11 | L76159_at | | 12.00 | 102.41 | 125.73 | 0 |

**Table 8.9** Fifteen models by four genes

| p | Y1 | Y2 | Y3 | Y4 | c | MNM | ZERO |
|---|----|----|----|----|---|-----|------|
| 4 | 1  | 1  | 1  | 1  | 1 | 0   | 0    |
| 3 | 1  | 1  | 1  | 0  | 1 | 1   | 0    |
|   | 1  | 1  | 0  | 1  | 1 | 1   | 0    |
|   | 1  | 0  | 1  | 1  | 1 | 3   | 0    |
|   | 0  | 1  | 1  | 1  | 1 | 2   | 0    |
| 2 | 1  | 1  | 0  | 0  | 1 | 2   | 0    |
|   | 1  | 0  | 1  | 0  | 1 | 4   | 0    |
|   | 0  | 1  | 1  | 0  | 1 | 3   | 0    |
|   | 1  | 0  | 0  | 1  | 1 | 4   | 0    |
|   | 0  | 1  | 0  | 1  | 1 | 13  | 0    |
|   | 0  | 0  | 1  | 1  | 1 | 6   | 0    |
| 1 | 1  | 0  | 0  | 0  | 1 | 5   | 0    |
|   | 0  | 1  | 0  | 0  | 1 | 25  | 0    |
|   | 0  | 0  | 1  | 0  | 1 | 10  | 0    |
|   | 0  | 0  | 0  | 1  | 1 | 17  | 0    |

15 models. Only the four-variable model is linearly separable. Therefore, we find one BGS, e.g., (Y1: M11722_at, Y2: X59871_at, Y3: U05259_rna1_at, Y4: D21063_at) in SM1. All MNMs, including these four genes, are linearly separable. Therefore, there are numerous linearly separable subspaces in the dataset. It is hard for us to analyze the dataset using ordinary statistical methods without knowing this fact.

If we directly search for the BGSs using all possible combination models of 11 variables in Eq. (8.1), it requires us to check 2047 models and has a huge computational time. Therefore, we propose a different idea in this section. In Table 8.8, because logistic regression tells us that the four-gene subspace is linearly separable in Eq. (8.2), we check only 15 models by all possible combinations in Table 8.9. Because the MNM of the four-gene model is zero and the four MNMs of the three-gene models are not zero, we know the four-gene model is the BGS. Therefore, we find the first Matroska series in Eq. (8.1); Golub11 is SM1. Next, we survey the smaller linearly separable model in SM1 by logistic regression in (8.2). Last, we confirm these four genes as the first BGS of SM1 in Eq. (8.3) by all combination models.

$$\text{Matroska series}: \text{Golub7, 129} \supseteq \text{Golub34} \supseteq \text{Golub11}. \quad (8.1)$$

$$\text{Logistic regression}: 11 \text{ genes} \rightarrow 4 \text{ genes}. \quad (8.2)$$

$$\text{Founding of BGSs}: 15 \text{ models} \rightarrow 4 \text{ genes}. \quad (8.3)$$

If we add another gene to the BGS, the 7125 (=7129 − 4) five-gene sets are linearly separable models because they include the four-gene subspace with

## 8.3 Results of the Golub et al. Dataset

MNM = 0. Therefore, there are numerous Matroskas from Golub7129 to Golub4 (BGS) in the dataset.

Until now, we have ignored the discrimination for the LSD. Some statisticians claim that discrimination is for overlapping data, not LSD. However, we can correctly define the status of "overlap" by the condition "MNM ≥ 1." Moreover, this knowledge is essential for understanding the dataset.

Although Vapnik clearly defined LSD by H-SVM, no researcher has discriminated the data using H-SVM. We believe there are two reasons:

1. We can discriminate only LSD by H-SVM. However, nobody uses H-SVM because it causes errors for data with "MNM ≥ 1." Moreover, we believe the discrimination of LSD is evident. However, there is no research on discrimination for LSD.
2. Because Vapnik proposed a fantastic kernel SVM, most researchers focus on the kernel SVM and ignore the importance of LSD in medical diagnosis, rating, pattern recognition, and gene analysis.

Because the LINGO program cannot find the BGSs directly, our research policy is as follows:

1. We find all SMs using the "LINGO Program 3" in Eq. (8.1).
2. Next, we search for BGSs by two stages in Eqs. (8.2) and (8.3).

### 8.3.3 Another BGS in the Fifth SM

We explain, again, how to find the BGSs in SM5. Although there are 13 genes in SM5, logistic regression finds the NM of seven genes is zero. Table 8.10 shows one Matroska with seven variables and two Matroskas with six variables with bold figures. If the model includes a variable, the value is "1". Otherwise, the value is "0". Although the MNM of a Matroska with seven variables is zero, only two Matroskas with six variables are zero. When we drop $Y1$ in the first Matroska with six variables, its MNM is 2. This fact means there are no BGSs in this subspace. On the other hand, if we drop $Y2$, the MNM of the second Matroska with six variables

**Table 8.10** Search for basic genes

| p | Y1 | Y2 | Y3 | Y4 | Y5 | Y6 | Y7 | C | MNM |
|---|----|----|----|----|----|----|----|----|-----|
| 7 | 1  | 1  | 1  | 1  | 1  | 1  | 1  | 1  | 0   |
| 6 | 0  | 1  | 1  | 1  | 1  | 1  | 1  | 1  | 2   |
| 6 | 1  | **0** | 1  | 1  | 1  | 1  | 1  | 1  | **0** |
| 6 | 1  | 1  | 0  | 1  | 1  | 1  | 1  | 1  | 1   |
| 6 | 1  | 1  | 1  | 0  | 1  | 1  | 1  | 1  | 1   |
| 6 | 1  | 1  | 1  | 1  | 0  | 1  | 1  | 1  | 1   |
| 6 | 1  | 1  | 1  | 1  | 1  | **0** | 1  | 1  | **0** |
| 6 | 1  | 1  | 1  | 1  | 1  | 1  | 0  | 1  | 2   |

is zero. This means there is/are BGS(s) in this Matroska with six variables. Although we can obtain clear results by all possible combinations, it requires extensive computation time if the number of variables exceeds ten. Now, we check the 63 models of the Swiss banknote data with six variables by all possible models. The MNM of the sixth six-variable model is also zero.

Table 8.11 shows the second stage of the procedure. The first row is the second six-variable model, dropping Y2 from the full model. We check five five-variable models from the second row to the seventh row. The MNM of the fifth five-variable model is zero. On the other hand, if we survey the sixth six-variable model without Y6, the MNM of this model is zero. Both surveys selected the same model, dropping Y2 and Y6 from the full model, e.g., (Y1, Y2, Y3, Y4, Y5, Y6, Y7) = (1, 0, 1, 1, 1, 0, 1).

Table 8.12 shows the stopping rule for this procedure. Because five MNMs of the four-variable models are greater than one, we can confirm the five-variable model is the BGS. We can summarize the procedure to find BGS as follows:

**Table 8.11** Search for basic gene sets

| p | Y1 | Y2 | Y3 | Y4 | Y5 | Y6 | Y7 | C | MNM |
|---|----|----|----|----|----|----|----|---|-----|
| 6 | 1 | 0 | 1 | 1 | 1 | 1 | 1 | 1 | 0 |
| 5 | 0 | 0 | 1 | 1 | 1 | 1 | 1 | 1 | 2 |
|   | 1 | 0 | 0 | 1 | 1 | 1 | 1 | 1 | 1 |
|   | 1 | 0 | 1 | 0 | 1 | 1 | 1 | 1 | 1 |
|   | 1 | 0 | 1 | 1 | 0 | 1 | 1 | 1 | 1 |
|   | 1 | 0 | 1 | 1 | 1 | 0 | 1 | 1 | 0 |
|   | 1 | 0 | 1 | 1 | 1 | 1 | 0 | 1 | 2 |
| 6 | 1 | 1 | 1 | 1 | 1 | 0 | 1 | 1 | 0 |
| 5 | 0 | 1 | 1 | 1 | 1 | 0 | 1 | 1 | 3 |
|   | 1 | 0 | 1 | 1 | 1 | 0 | 1 | 1 | 0 |
|   | 1 | 1 | 0 | 1 | 1 | 0 | 1 | 1 | 1 |
|   | 1 | 1 | 1 | 0 | 1 | 0 | 1 | 1 | 1 |
|   | 1 | 1 | 1 | 1 | 0 | 0 | 1 | 1 | 1 |
|   | 1 | 1 | 1 | 1 | 1 | 0 | 0 | 1 | 2 |

**Table 8.12** Stopping rule

| p | Y1 | Y2 | Y3 | Y4 | Y5 | Y6 | Y7 | C | IC |
|---|----|----|----|----|----|----|----|---|----|
| 5 | 1 | 0 | 1 | 1 | 1 | 0 | 1 | 1 | 0 |
| 4 | 0 | 0 | 1 | 1 | 1 | 0 | 1 | 1 | 3 |
|   | 1 | 0 | 0 | 1 | 1 | 0 | 1 | 1 | 1 |
|   | 1 | 0 | 1 | 0 | 1 | 0 | 1 | 1 | 1 |
|   | 1 | 0 | 1 | 1 | 0 | 0 | 1 | 1 | 1 |
|   | 1 | 0 | 1 | 1 | 1 | 0 | 0 | 1 | 4 |

8.3 Results of the Golub et al. Dataset

1. We find two Matroska series in SM5, e.g., (1, 1, 1, 1, 1, 1, 1) ⊇ (1, 0, 1, 1, 1, 1, 1) ⊇ (1, 0, 1, 1, 1, 0, 1) and (1, 1, 1, 1, 1, 1, 1) ⊇ (1, 1, 1, 1, 1, 0, 1) ⊇ (1, 0, 1, 1, 1, 0, 1).
2. By the stopping rule, the five-gene subspace ($Y1$, $Y3$–$Y4$, $Y7$) is BGS.

## 8.4 How to Analyze the First BGS

Figure 8.1 shows the cluster analysis of BGS in SM1 by the Ward method. Although all output of case cluster is omitted, we find four cases belonging to the acute lymphoblastic leukemia (ALL) class, e.g., 42, 63, 67, and 62, enter the acute myeloid leukemia (AML) cluster. The cluster of variables tells us that "M11722_at" and "U05259_rna1_at" become the first cluster. Next, "D21063_at" joins the first cluster and becomes the second cluster. At last, "X59871_at" joins the second cluster, and the clustering ends.

Figure 8.2 shows the result of PCA. The left figure is the eigenvalues. Two eigenvalues are greater than one, and the contribution ratio is about 0.75. The middle figure is the scatter plot. AMLs are located in the third quadrant. 47 ALL cases are situated in the fourth quadrant, first quadrant, and second quadrant. The right plot is the factor-loading plot. "M11722_at" is overlapped on the X-axis, and "X59871_at" is overlapped on the Y-axis. Figure 8.3 shows two score plots.

**Fig. 8.1** Part of cluster analysis

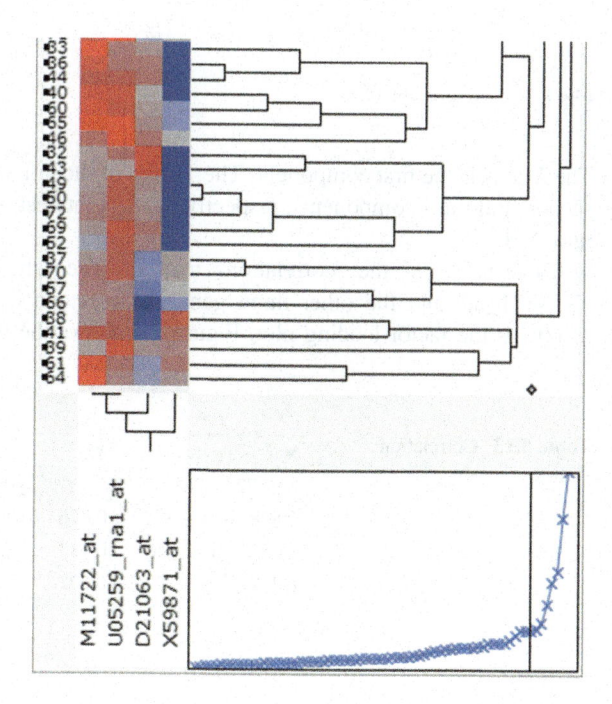

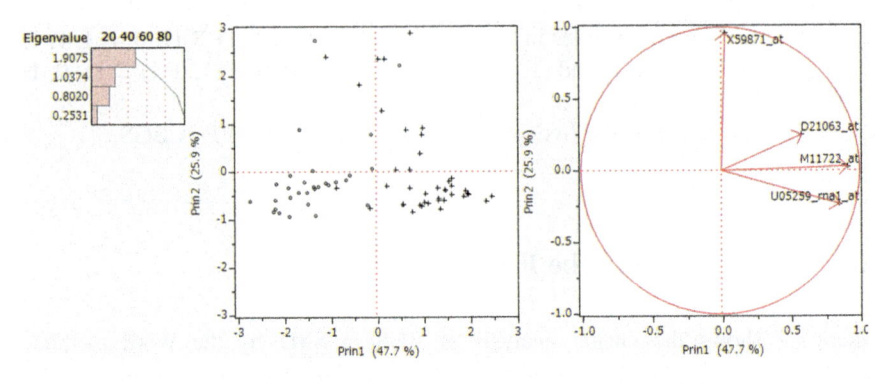

**Fig. 8.2** PCA (*left* eigenvalues, *Middle* scatter plot, *right* factor-loading plot)

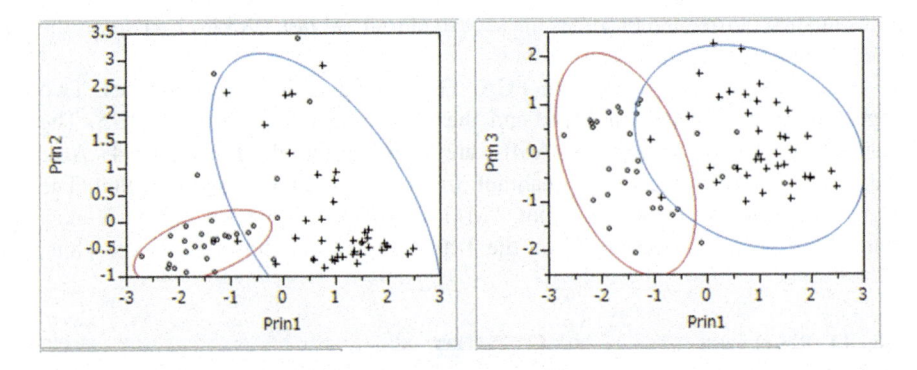

**Fig. 8.3** Two score plots

The *X*-axis is the first component. The *Y*-axes of the left and right score plots are the second and third components, respectively. It is interesting that AML seems to bite into ALL.

Table 8.13 is the correlation matrix. The absolute correlation among "X59871_at" and the other three genes is less than 0.088. Those are the same results as the factor-loading plot. Figure 8.4 shows the matrix scatter plot.

**Table 8.13** Correlation matrix

| Var. | X1 | X2 | X3 | X4 |
|---|---|---|---|---|
| M11722_at | 1 | 0.076 | 0.713 | 0.371 |
| X59871_at | 0.076 | 1 | −0.088 | 0.052 |
| U05259_rna1_at | 0.713 | −0.088 | 1 | 0.220 |
| D21063_at | 0.371 | 0.052 | 0.220 | 1 |

## 8.5 Statistical Analysis of SM1

**Fig. 8.4** Matrix scatter plot

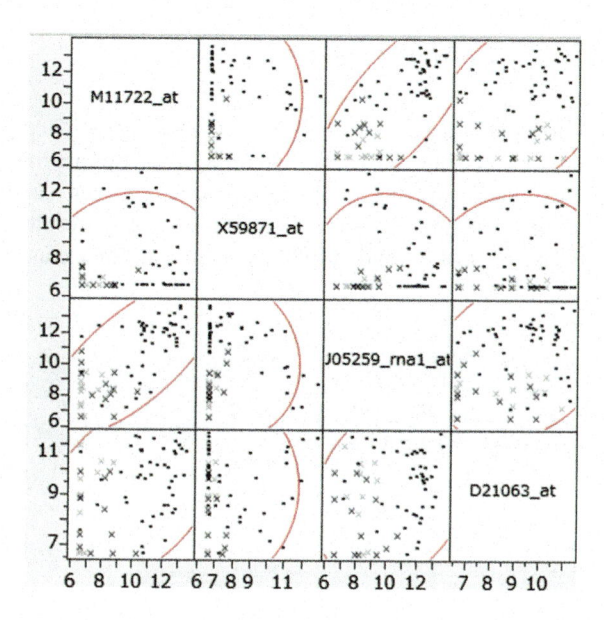

## 8.5 Statistical Analysis of SM1

In this section, we analyze SM1 and compare four genes in BGS and another seven genes by a statistical method. The four genes are "D21063_at, M11722_at, U05259_rna1_at, and X59871_at" and the seven genes are "X883, X1304, X1640, X1809, X1828, X1856, and X4680." The two types of genes have different variable names, showing the differences of the two groups. Many researchers analyze the dataset by ordinary statistical methods, such as one-way ANOVA, cluster analysis, PCA, and regression analysis. We fear those methods do not provide clear results for the dataset without considering the Matroska structure.

### 8.5.1 One-Way ANOVA

Figure 8.5 shows the one-way ANOVA of four genes in BGS and another seven genes. The t-values of the four genes are 3.843, 10.231, 8.396, and 3.429. The minimum value is 3.429. Those of the seven genes are −0.813, −0.698, 2.448, 1.576, −0.966, −1.407, and 6.727. Because six absolute $t$-values, except for X4680, are less than 3.429, four genes in BGS are important for the discrimination of two classes. However, because the $t$-value of X4680 is 6.727, we focus on this gene. Although four-variable model and 11-variable model are linearly separable, 11 box–whisker plots are overlapped. This fact tells us the difficulties of the feature-selection using ordinary statistical methods such as one-way ANOVA, cluster analysis and PCA.

## 8.5.2 Cluster Analysis

When the Ward cluster analysis method analyzes 72 cases with 11 genes and makes two clusters, one AML case and nine ALL cases are misclassified to another cluster. This fact explains why the dataset cluster analysis is not helpful for feature-selection. Figure 8.6 shows the result of variable clustering. (U05259_rna1_at and X4680) becomes the first cluster. (D21063_at and X1640)

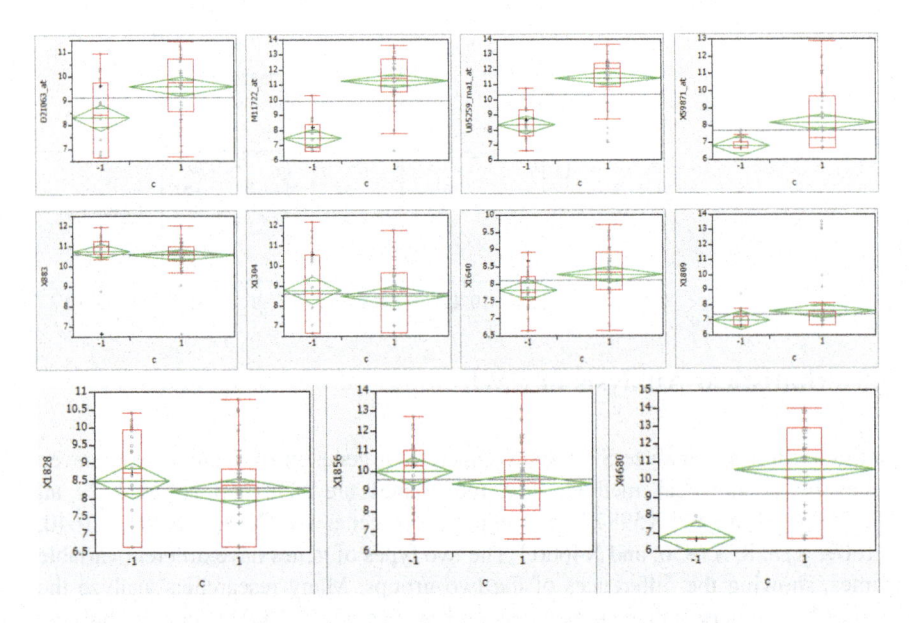

**Fig. 8.5** One-way ANOVA (*Upper* four genes in BGS; *middle* and *low* seven genes)

**Fig. 8.6** Cluster analysis (ward method)

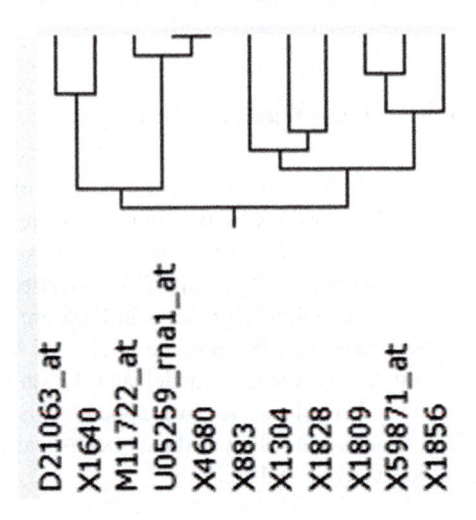

8.5 Statistical Analysis of SM1   185

becomes the fourth cluster, and (X59871_at and X1809) becomes the third cluster. It is very interesting to us that three genes included in the BGS are clustered with three other genes not included in the BGS.

### 8.5.3 PCA

Figure 8.7 shows the PCA. The left figure is the eigenvalues. Four eigenvalues are greater than one, and the contribution ratio is about 0.68. The middle figure is the scatter plot. AMLs are located in the second and third quadrant. 47 ALL cases are situated in the first and fourth quadrant. The right figure is the factor-loading plot. "M11722_at" is overlapped on the $X$-axis and "X59871_at" is overlapped on the $Y$-axis.

Figure 8.8 shows scatter plot. The $X$-axis is the first component. The $Y$-axis is the second component. Because each ellipse include both class cases, PCA may not be helpful for gene analysis.

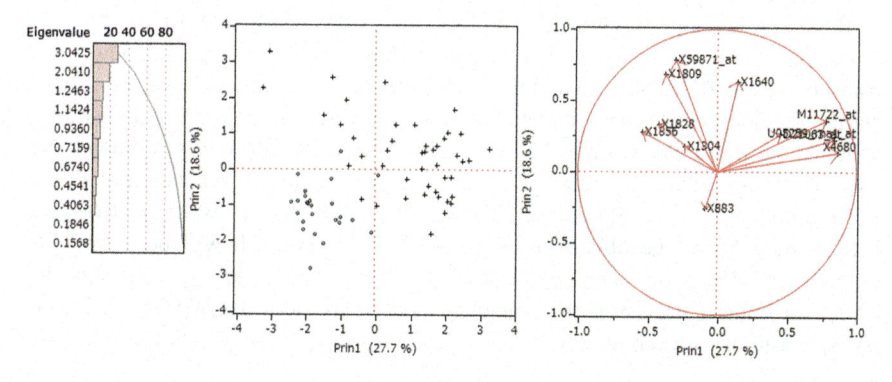

**Fig. 8.7** PCA

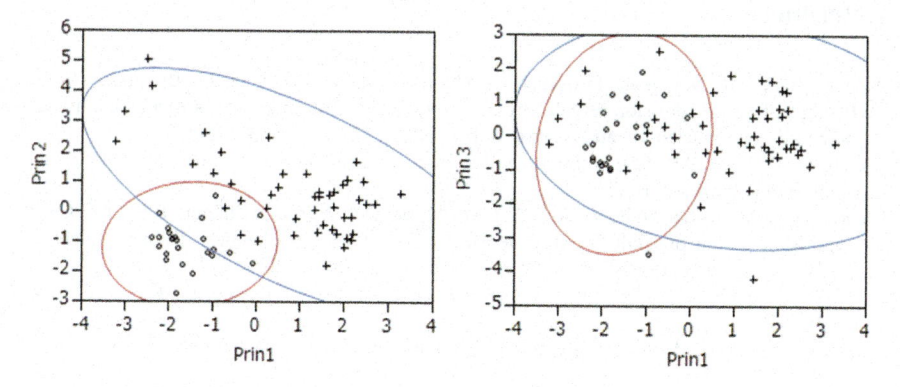

**Fig. 8.8** Scatter plot

## 8.6 Summary

We developed the Theory, Method 1, and Method 2. Revised IP-OLDF solves Problem 1, Problem 2, and Problem 5. Moreover, the best models of Revised IP-OLD are better than another seven LDFs. Although H-SVM correctly discriminates LSD, it cannot select features for six dataset. Because Problem 3 is a defect of the generalized inverse matrices technique and the error rates of Fisher's LDF and QDF are very high for LSD in Chap. 5, we believe that discriminant analysis and regression analysis based on variance–covariance matrices may not be helpful for gene analysis. Although the discriminant analysis is not an inferential statistical method, the Method 1 offers the 95 % CI for the error rate and the discriminant coefficient, and the validation of Revised IP-OLDF by six different types of data.

In this paper, we do not discuss the validation of our results. However, because the Method 1 already validates six ordinary data, the validation of dataset does not matter. Because the best model is a powerful feature-selection procedure for ordinary data, we ignore that some parameters of Revised IP-OLDF become zero in ordinary data. Because other LDFs cannot naturally select features, they may be difficult for gene dataset.

If we can develop "Revised LINGO Program 3" that can find all BGSs, it will be more useful in gene analysis. The "LINGO Program 3" is useful for other gene dataset, such as RNA-Seq., in addition to the six datasets. Although we surveyed to clarify the long-term survivors of the Maruyama vaccine (SSM) administration patients, our trial failed (Shinmuea et al. 1987; Shinmura 2001). If we compare two lists of cancer genes, (normal and cancer patient dataset) versus (normal and SSM-administration patient dataset) and find the differences between the two gene lists, it may prove the effectiveness of SSM. This approach will be helpful for judging the effects of other cancer treatments, except for surgery. We hope for advice or dataset provision from medical doctors or projects. We would like to propose a joint research study.

## References

Alon U, Barkai N, Notterman DA, Gish K, Ybarra S, Mack D, Levine AJ, Broad AJ (1999) Patterns of gene expression revealed by clustering analysis of tumor and normal colon tissues probed by oligonucleotide arrays. Proc Natl Acad Sci USA 96:6745–6750

Buhlmann P, Geer AB (2011) Statistics for high-dimensional dataset method, theory and applications. Springer, Berlin

Chiaretti S, Li X, Gentleman R, Vitale A, Vignetti M, Mandelli F, Ritz J, Chiaretti RF (2004) Gene expression profile of adult T-cell acute lymphocytic leukemia identifies distinct subsets of patients with different response to therapy and survival. Blood, 103/7:2771–2778

Cox DR (1958) The regression analysis of binary sequences (with discussion). J Roy Stat Soc B 20:215–242

Firth D (1993) Bias reduction of maximum likelihood estimates. Biometrika 80:27–39

# References

Fisher RA (1936) The use of multiple measurements in taxonomic problems. Ann Eugenics 7:179–188

Fisher RA (1956) Statistical methods and statistical inference. Hafner Publishing Co, New Zealand

Flury B, Rieduyl H (1988) Multivariate statistics: a practical approach. Cambridge University Press, Cambridge

Friedman JH (1989) Regularized discriminant analysis. J Am Stat Assoc 84(405):165–175

Golub TR, Slonim DK, Tamayo P, Huard C, Gaasenbeek M, Mesirov JP, Coller H, Loh ML, Downing JR, Caligiuri MA, Bloomfield CD, Lander ES (1999) Molecular Classification of Cancer: Class Discovery and Class Prediction by Gene Expression Monitoring. Science 286 (5439):531–537

Goodnight JH (1978) SAS technical report—the sweep operator: its importance in statistical computing—(R100). SAS Institute Inc, USA

Jeffery IB, Higgins DG, Culhane C (2006) Comparison and evaluation of methods for generating differentially expressed gene lists from microarray data. BMC Bioinf. Jul 26 7:359:1–16. doi:10.1186/1471-2105-7-359

Lachenbruch PA, Mickey MR (1968) Estimation of error rates in discriminant analysis. Technometrics 10:1–11

Miyake A, Shinmura S (1976) Error rate of linear discriminant function. In: Dombal FT, Gremy F (ed). North-Holland Publishing Company, The Netherland, pp 435–445

Sall JP, Creighton L, Lehman A (2004) JMP start statistics, third edition. SAS Institute Inc., USA. (Shinmura S. edits Japanese version)

Schrage L (2006) Optimization modeling with LINGO. LINDO Systems Inc., USA. (Shinmura S. translates Japanese version)

Shinmura S, Iida K, Maruyama C (1987) Estimation of the effectiveness of cancer treatment by SSM using a null hypothesis model. Inf Health Social Care 7(3):263–275. doi:10.3109/14639238709010089

Shinmura S (1998) Optimal linear discriminant functions using mathematical programming. J Jpn Soc Comput Stat, 11/2: 89–101

Shinmura S, Tarumi T (2000) Evaluation of the optimal linear discriminant functions using integer programming (ip-oldf) for the normal random dataset. J Jpn Soc Comput Stat 12(2):107–123

Shinmura S (2000a) A new algorithm of the linear discriminant function using integer programming. New Trends Prob Stat 5:133–142

Shinmura S (2000b) Optimal linear discriminant function using mathematical programming. Dissertation, March 200:1–101, Okayama University, Japan

Shinmura S (2001) Analysis of effect of SSM on 152,989 cancer patient. ISI2001, pp 1–2. doi:10.13140/RG.2.1.30779281

Shinmura S (2003) Enhanced algorithm of IP-OLDF. ISI2003 CD-ROM, pp 428–429

Shinmura S (2004) New algorithm of discriminant analysis using integer programming. IPSI 2004 Pescara VIP Conference CD-ROM, pp 1–18

Shinmura S (2005) New age of discriminant analysis by IP-OLDF—beyond Fisher's linear discriminant function. ISI2005, pp 1–2

Shinmura S (2007) Overviews of discriminant function by mathematical programming. J Jpn Soc Comput Stat 20(1–2):59–94

Shinmura S (2010a) The optimal linearly discriminant function (Saiteki Senkei Hanbetu Kansuu). Union of Japanese Scientist and Engineer Publishing, Japan

Shinmura S (2010b) Improvement of CPU time of Revised IP-OLDF using Linear Programming. J Jpn Soc Comput Stat 22(1):39–57

Shinmura S (2011a) Beyond Fisher's linear discriminant analysi—new world of the discriminant analysis. ISI CD-ROM, pp 1–6

Shinmura S (2011b) Problems of discriminant analysis by mark sense test data. Jpn Soc Appl Stat 40(3):157–172

Shinmura S (2013) Evaluation of optimal linear discriminant function by 100-fold cross-validation. ISI CD-ROM, pp 1–6

Shinmura S (2014a) End of discriminant functions based on variancE-covariance matrices. ICORE2014, pp 5–16

Shinmura S (2014b) Improvement of CPU time of linear discriminant functions based on MNM criterion by IP. Stat Optim Inf Comput 2:114–129

Shinmura S (2014c) Comparison of linear discriminant functions by $k$-fold cross-validation. Data Anal 2014:1–6

Shinmura S (2015a) The 95 % confidence intervals of error rates and discriminant coefficients. Stat Optim Inf Comput 2:66–78

Shinmura S (2015b) A trivial linear discriminant function. Stat Optim Inf Comput 3:322–335. doi:10.19139/soic.20151202

Shinmura S (2015c) Four serious problems and new facts of the discriminant analysis. In: Pinson E, Valente F, Vitoriano B (ed) Operations research and enterprise systems, pp 15–30. Springer, Berlin (ISSN: 1865-0929, ISBN: 978-3-319-17508-9, doi:10.1007/978-3-319-17509-6)

Shinmura S (2015d) Four problems of the discriminant analysis. ISI 2015:1–6

Shinmura S (2015e) The discrimination of microarray data (Ver. 1). Res Gate 1:1–4. 28 Oct 2015

Shinmura S (2015f) Feature selection of three microarray data. Res Gate 2:1–7. 1 Nov 2015

Shinmura S (2015g) Feature Selection of Microarray Data (3)—Shipp et al. Microarray Data. Res Gate (3) 1–11

Shinmura S (2015h) Validation of feature selection (4)—Alon et al. microarray data. Res Gate (4) 1–11

Shinmura S (2015i) Repeated feature selection method for microarray data (5). Res Gate 5:1–12. 9 Nov 2015

Shinmura S (2015j) Comparison Fisher's LDF by JMP and revised IP-OLDF by LINGO for microarray data (6). Res Gate 6:1–10. 11 Nov 2015

Shinmura S (2015k) Matroska trap of feature selection method (7)—Golub et al. microarray data. Res Gate (7), 18:1–14

Shinmura S (2015l) Minimum Sets of Genes of Golub et al. Microarray Data (8). Research Gate (8) 1–12. 22 Nov 2015

Shinmura S (2015m) Complete lists of small matroska in Shipp et al. microarray data (9). Res Gate (9) 1–81

Shinmura S (2015n) Sixty-nine small matroska in Golub et al. microarray data (10). Res Gate 1–58

Shinmura S (2015o) Simple structure of Alon et al. microarray data (11). Res Gate(1.1) 1–34

Shinmura S (2015p) Feature selection of Singh et al. microarray data (12). Res Gate (12) 1–89

Shinmura S (2015q) Final list of small matroska in Tian et al. microarray data. Res Gate (13) 1–160

Shinmura S (2015r) Final list of small matroska in Chiaretti et al. microarray data. Res Gate (14) 1–16

Shinmura S (2015s) Matroska feature selection method for microarray dataset. Res Gate (15) 1–16

Shinmura S (2016a) The best model of swiss banknote data. Stat Optim Inf Comput, 4:118–131. International Academic Press (ISSN: 2310-5070 (online) ISSN: 2311-004X (print), doi:10.19139/soic.v4i2.178)

Shinmura S (2016b) Matroska feature–selection method for microarray data. Biotechnology 2016:1–6

Shinmura S (2016c) Discriminant analysis of the linear separable dataset—Japanese automobiles. J Stat Sci Appl X, X:0–14

Shipp MA, Ross KN, Tamayo P, Weng AP, Kutok JL, Aguiar RC, Gaasenbeek M, Angelo M, Reich M, Pinkus GS, Ray TS, Koval MA, Last KW, Norton A, Lister TA, Mesirov J, Neuberg DS, Lander ES, Aster JC, Golub TR (2002) Diffuse large B-cell lymphoma outcome prediction by gene-expression profiling and supervised machine learning. Nat Med 8:68–74

Simon N, Friedman J, Hastie T, Tibshirani R (2013) A sparse-group lasso. J Comput Graph Stat 22:231–245

# References

Singh D, Febbo PG, Ross K, Jackson DG, Manola J, Ladd C, Tamayo P, Renshaw AA, D'Amico A, Richie JP, Lander ES, Lada M, Kantoff PW, Golub TR, Sellers WR (2002) Gene expression correlates of clinical prostate cancer behavior. Cancer Cell 1:203–209

Stam A (1997) Non-traditional approaches to statistical classification: some perspectives on Lp-norm methods. Ann Oper Res 74:1–36

Tian E, Zhan F, Walker R, Rasmussen E, Ma Y, Barlogie B, Shaughnessy JD (2003) The role of the Wnt-signaling antagonist DKK1 in the development of osteolytic lesions in multiple myeloma. New Engl J Med 349(26):2483–2494

Vapnik V (1995) The nature of statistical learning theory. Springer, Berlin

# Chapter 9
# LINGO Program 2 of Method 1

## 9.1 Introduction

Although Fisher established the statistical discriminant analysis based on Fisher's assumption, he did not define SE of error rate and discriminant coefficient. Therefore, we proposed the 100-fold cross-validation for small sample method (the Method 1). We develop LINGO Program 2 of Method 1 and explain it in this section.[1] The Method 1 is the combination of resampling and $k$-fold cross-validation. The Method 1 is as follows: (1) We copy 100 times the data as validation sample from the original data using JMP. (2) We add a uniform random number as a new variable, sort the data in ascending order, and divide into 100 subsets as 100 training samples. (3) We evaluate eight LDFs by the Method 1 using these 100 subsets as training samples and unique validation sample. I analyze six MP-based LDFs by LINGO, developed with the support of LINDO Systems Inc. I analyze logistic regression and Fisher's LDF by JMP, obtained with the assistance of the JMP division of SAS Japan. There is merit in using 100-fold cross-validation because we can easily calculate the 95 % CI of the discriminant coefficients and error rates. Moreover, error rate means, $M1$ and $M2$, in the training and validation samples offer direct and powerful model selection procedure such as the best model. We can show the best models of Revised IP-OLDF are better than other LDFs. We can use the LOO procedure for model selection of discriminant analysis but cannot obtain the 95 % CI. These differences are quite important for the analysis of small samples.

---

[1]Manuals and LINDO products such as LINGO, What's Best! (Excel add-in solver), and LINDO API (C library) are free downloaded from LINDO Systems Inc. (http://www.lindo.com/).

© Springer Science+Business Media Singapore 2016
S. Shinmura, *New Theory of Discriminant Analysis After R. Fisher*,
DOI 10.1007/978-981-10-2164-0_9

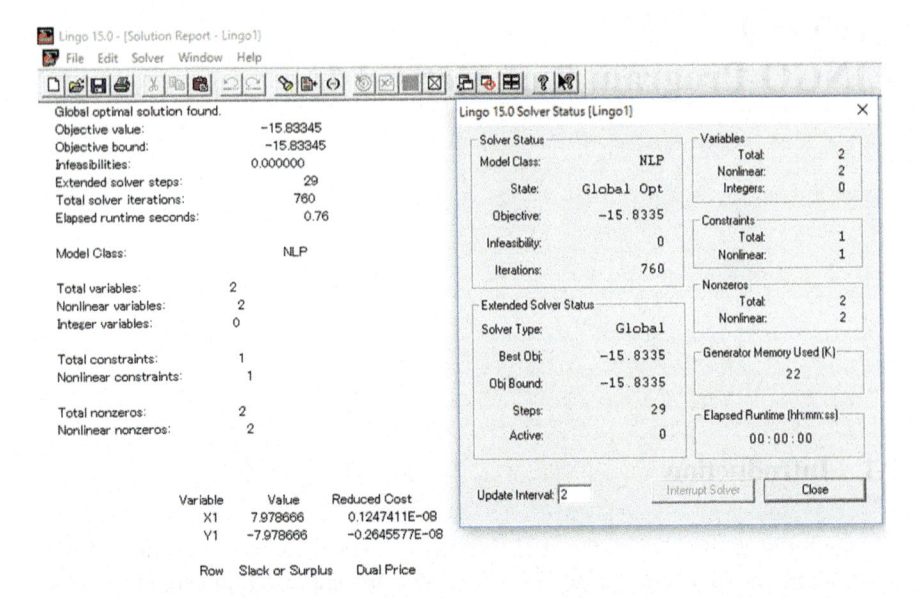

**Fig. 9.1** Global minimum solution

## 9.2 Natural (Mathematical) Notation by LINGO

In this section, we explain the natural (mathematical) notation of Revised IP-OLDF by LINGO. We can describe the model by two notations such as "natural notation" and "SET and CALC" notation." Natural notation is similar to the usual mathematical notation. In the case where we want to obtain the global minimum value of the function[2] $Z(x, y) = x \times \sin(y) + y \times \sin(x)$, we define the object function and two constraints and obtain the output as shown in Fig. 9.1. The symbol "@" represents the LINGO function. "@BND (−10, X1, 10)" means the two constraints (−10 ≤ X1 ≤ 10). Within 1 s, we obtain −15.8 as the global minimum solution (the minimum value, not local minimum) at (x1, y1) = (7.98, −7.98).

```
MIN=X1*@SIN(Y1)+Y1*@SIN(X1);
@BND(-10,X1,10); @BND(-10,Y1,10);
```

---

[2]Three LINDO solvers can offer global solution (maximum and minimum values) and is useful in mathematical education and many researches.

## 9.2 Natural (Mathematical) Notation by LINGO

**Fig. 9.2** Surface graph

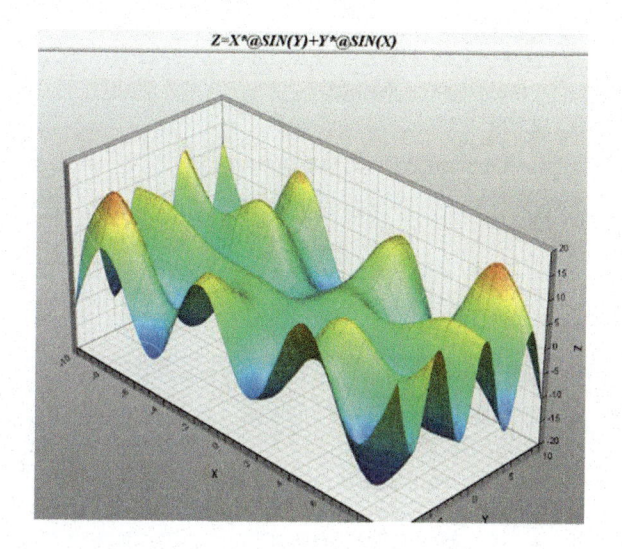

$Z=X*@SIN(Y)+Y*@SIN(X)$

If we want to draw a contour or surface graph, we can define the model with the "SET and CALC" section. The "MODEL:" section consists of two subsections, such as "SETS and CALC" and ends with "END." We insert the "SETS: ... ENDSETS" and "CALC: ... ENDCALC" sections before and after the optimization model. The "SETS" section defines a one-dimensional set, such as "POINTS," with 21 elements by "/1...21/." We can define two-dimensional sets by the combination of the one-dimensional set, such as "POINT2 (POINTS, POINTS):" that is a two-dimensional set with (21, 21) elements. We can define three-dimensional sets by the combination of three one-dimensional sets. "POINT2" defines three arrays, such as "$X, Y, Z$" with (21, 21) elements on the right side of ":". The set "POINTS" has no one-dimensional array. The "CALC" section is a programming language that can optimize the MP-model and control it. In the "CALC" section, we draw the surface graph of $Z(x, y)$ at the mesh by $x = (-10, -9, ..., 9, 10)$ and $y = (-10, -9, ..., 9, 10)$. Figure 9.2 shows the surface graph. Because XS = @FLOOR $(-(@SIZE (POINTS)/2) + 0.5) =$ @FLOOR $(-(21/2) + 0.5) =$ @FLOOR $(-10) = -10$, we can directly replace this statement by "$X = -10$". "@FOR (POINTS2(I, J):" is the below loop function. Because the MP-based model has the same structure of constraints, we need not describe those constraints one by one. The "@CHART (..., 'SURFACE (or CONTOUR)',...);" function draws the surface (or contour) shown in Fig. 9.2. If we replace "surface" with "contour," we can draw the contour graph.

```
For i = 1 To 21
  For j To 1 Step 21
Xij=XS+I-1;
Yij=YS+J-1;
Zij=Xij*SIN(Yij)+Yij*SIN(Xij);
  Next j
Next i
```

```
MODEL:
SETS:
 POINTS /1..21/;
 POINTS2 (POINTS,POINTS): X,Y,Z;
ENDSETS

MIN=X1*@SIN(Y1)+Y1*@SIN(X1);
@BND(-10,X1,10); @BND(-10,Y1,10);

CALC:
XS=@FLOOR(-(@SIZE(POINTS)/2)+0.5);
YS=XS;
@FOR (POINTS2(I,J):
 X(I,J)=XS+ I-1;
 Y(I,J)=YS+ J-1; Z(I,J)=X(I,J)*@SIN(Y(I,J))+Y(I,J)*@SIN(X(I,J)));
@CHART( 'X Y Z', 'SURFACE', 'Z = X*@SIN(Y) +Y*@SIN(X)' );
ENDCALC
END
```

## 9.3 Iris Data in Excel

We discriminate the Iris data in an Excel file by Revised IP-OLDF. These data consist of two species, such as versicolor ($y_i = 1$) and virginica ($y_i = -1$), as shown in Fig. 9.3. Each species has 50 cases with five variables (four independent variables and indicator (object variable) $y_i$). We define the Excel range name "IS" as "B2: F101." LINGO can retrieve "IS" array values by the "IS = @ OLE();" function and use it as the LINGO array name "IS."

## 9.3 Iris Data in Excel

**Fig. 9.3** Iris data in Excel

| | A | B | C | D | E | F |
|---|---|---|---|---|---|---|
| 1 | 種類 | X1 | X2 | X3 | X4 | y |
| 2 | versicolor | 7 | 3.2 | 4.7 | 1.4 | 1 |
| 3 | versicolor | 6.4 | 3.2 | 4.5 | 1.5 | 1 |
| 4 | versicolor | 6.9 | 3.1 | 4.9 | 1.5 | 1 |
| 5 | versicolor | 5.5 | 2.3 | 4 | 1.3 | 1 |
| 6 | versicolor | 6.5 | 2.8 | 4.6 | 1.5 | 1 |
| 7 | versicolor | 5.7 | 2.8 | 4.5 | 1.3 | 1 |
| 8 | versicolor | 6.3 | 3.3 | 4.7 | 1.6 | 1 |
| 9 | versicolor | 4.9 | 2.4 | 3.3 | 1 | 1 |
| 10 | versicolor | 6.6 | 2.9 | 4.6 | 1.3 | 1 |
| 11 | versicolor | 5.2 | 2.7 | 3.9 | 1.4 | 1 |
| 12 | versicolor | 5 | 2 | 3.5 | 1 | 1 |
| 13 | versicolor | 5.9 | 3 | 4.2 | 1.5 | 1 |
| 14 | versicolor | 6 | 2.2 | 4 | 1 | 1 |
| 15 | versicolor | 6.1 | 2.9 | 4.7 | 1.4 | 1 |
| 16 | versicolor | 5.6 | 2.9 | 3.6 | 1.3 | 1 |
| 50 | versicolor | 5.1 | 2.5 | 3 | 1.1 | 1 |
| 51 | versicolor | 5.7 | 2.8 | 4.1 | 1.3 | 1 |
| 52 | virginica | −6.3 | −3.3 | −6 | −2.5 | −1 |
| 53 | virginica | −5.8 | −2.7 | −5.1 | −1.9 | −1 |
| 100 | virginica | −6.2 | −3.4 | −5.4 | −2.3 | −1 |
| 101 | virginica | −5.9 | −3 | −5.1 | −1.8 | −1 |

Next, we define the Excel range name "CHOICE" as "H2: L16." A total of 15 rows correspond to the models from the full model ($X1$, $X2$, $X3$, $X4$) to the one-variable model ($X2$). After optimization, we output three LINGO arrays by the "@OLE() = IC, ZERO, VARK100;" function as shown in Figs. 9.4 and 9.5. "IC" includes the "NM" in the Excel array name "NM ($M2$: $M16$)" in Fig. 9.4. "ZERO" includes the number of discriminant hyperplanes in the LINGO array name "ZERO ($N2$:$N16$)." "VARK100" includes the discriminant coefficients in "VARK100 ($O2$: $S16$)" in Fig. 9.5.

**Fig. 9.4** Range name CHOICE, NM, and ZERO

| | G | H | I | J | K | L | M | N |
|---|---|---|---|---|---|---|---|---|
| 1 | | x1 | x2 | x3 | x4 | c | NM | ZERO |
| 2 | 1 | 1 | 1 | 1 | 1 | 1 | 1 | 0 |
| 3 | 2 | 0 | 1 | 1 | 1 | 1 | 2 | 0 |
| 4 | 3 | 1 | 0 | 1 | 1 | 1 | 2 | 0 |
| 5 | 4 | 1 | 1 | 0 | 1 | 1 | 4 | 0 |
| 6 | 5 | 1 | 1 | 1 | 0 | 1 | 2 | 0 |
| 7 | 6 | 0 | 1 | 0 | 1 | 1 | 5 | 0 |
| 8 | 7 | 0 | 0 | 1 | 1 | 1 | 3 | 0 |
| 9 | 8 | 1 | 0 | 1 | 0 | 1 | 4 | 0 |
| 10 | 9 | 1 | 0 | 0 | 1 | 1 | 5 | 0 |
| 11 | 10 | 0 | 1 | 1 | 0 | 1 | 6 | 0 |
| 12 | 11 | 1 | 1 | 0 | 0 | 1 | 25 | 0 |
| 13 | 12 | 0 | 0 | 0 | 1 | 1 | 6 | 0 |
| 14 | 13 | 0 | 0 | 1 | 0 | 1 | 7 | 0 |
| 15 | 14 | 1 | 0 | 0 | 0 | 1 | 27 | 0 |
| 16 | 15 | 0 | 1 | 0 | 0 | 1 | 37 | 0 |

**Fig. 9.5** Total of 15 Revised IP-OLDFs

|    | O       | P      | Q        | R       | S       |
|----|---------|--------|----------|---------|---------|
|    | x1      | x2     | x3       | x4      | c       |
| 1  | 6.9     | 13.4   | -30.1    | -32.2   | 120.3   |
| 2  | 0.0     | 20.0   | -16.0    | -52.0   | 109.4   |
| 3  | 7.4     | 0.0    | -14.4    | -11.9   | 43.8    |
| 4  | -1.5    | 12.3   | 0.0      | -42.6   | 47.3    |
| 5  | 6673.3  | 9980.0 | -23346.7 | 0.0     | 42083.0 |
| 6  | 0.0     | 50.0   | 0.0      | -160.0  | 129.0   |
| 7  | 0.0     | 0.0    | -19992.0 | -40004.0| 157969.8|
| 8  | 33.3    | 0.0    | -86.7    | 0.0     | 211.0   |
| 9  | 1.1     | 0.0    | 0.0      | -26.7   | 38.9    |
| 10 | 0.0     | 5882.4 | -23529.4 | 0.0     | 94118.6 |
| 11 | -40.0   | -10.0  | 0.0      | 0.0     | 278.0   |
| 12 | 0.0     | 0.0    | 0.0      | -20.0   | 33.0    |
| 13 | 0.0     | 0.0    | -20.0    | 0.0     | 97.0    |
| 14 | -8331.7 | 0.0    | 0.0      | 0.0     | 50824.2 |
| 15 | 0.0     | -20.0  | 0.0      | 0.0     | 59.0    |

## 9.4 Six LDFs by LINGO

We explain the Method 1 by LINGO, which is the solver developed by LINDO Systems Inc. I develop six LDFs by the "SET" notation: Revised IP-OLDF (RIP), Revised IPLP-OLDF (IPLP), Revised LP-OLDF (LP), H-SVM, and two S-SVMs (SVM4 and SVM1). We consider that the two S-SVMs are different LDFs. Revised IPLP-OLDF is the two-stage algorithm of Revised LP-OLDF and Revised IP-OLDF. The Revised IP-OLDF shown in Eq. (9.1) can find the actual MNM by "MIN = $\Sigma e_i$" because it can directly find the OCP interior point. If case $\mathbf{x_i}$ is classified, $e_i = 0$. If case $\mathbf{x_i}$ is misclassified, $e_i = 1$. Because the discriminant score becomes negative for the misclassified case, Revised IP-OLDF selects an alternative support vector (SV), such as "$y_i \times ({}^t\mathbf{x_i}\,\mathbf{b} + b_0) = 1 - M \times e_i = -9999$" instead of "$y_i \times ({}^t\mathbf{x_i}\,\mathbf{b} + b_0) = 1$" for classified cases.

$$\text{MIN} = \Sigma e_i; \; y_i \times ({}^t\mathbf{x_i}\mathbf{b} + b_0) \geq 1 - M \times e_i \tag{9.1}$$

**b**: $p$ independent variables, $b_0$: intercept
$\mathbf{x_i}$: $(1 \times p)$ case vector if data is $(n \times p)$
$({}^t\mathbf{x_i}\,\mathbf{b} + b_0)$: discriminant score
$M$: large $M$ constant, such as 10,000
$y_i$: $y_i = 1$ for class 1 and $y_i = -1$ for class 2
$e_i$: 0/1 integer decision variable that corresponds to $\mathbf{x_i}$

We can define this model in the "SUBMODEL" section of LINGO. "RIP" is the submodel name of Revised IP-OLDF. We can solve and control this IP model by this name. "@SUM and @FOR" are two important LINGO loop functions. "@SUM (N (i): E(i))" means "$\Sigma_{i=1}^{n} E(i)$." "@FOR(N(i):" defines n constraints such as "@SUM(P1 (j):IS(i, j) * VARK(j) * CHOICE(k, j)) ≥ 1 - BIGM * E(i))." "@FOR(P1(j): @FREE(VARK(j)));" defines **b** as the free decision variable for j = 1,..., (p+1). "@FOR(N(i):@BIN(E(i)));" defines "$e_i$" as the 0/1 integer decision variable for i = 1,...,n.

## 9.4 Six LDFs by LINGO

```
SUBMODEL RIP:
 MIN=ER;   ER=@SUM(N(i):E(i));
 @FOR(N(i):@SUM(P1(j):IS(i,j)*VARK(j)*CHOICE(k,j))
    >= 1-BIGM*E(i));
 @FOR(P1(j):@FREE(VARK(j)));
 @FOR(N(i):@BIN(E(i)));
ENDSUBMODEL
```

If we insert "!" before "@FOR($N(i)$:@BIN($E(i)$));", it changes the comment and "$e_i$" becomes a non-negative real decision variable from 0/1 binary integer. This model is Revised LP-OLDF. The "SUBMODEL" section defines an arbitrary character string. Therefore, we define the model of Revised LP-OLDF called "LP."

```
SUBMODEL LP:
 MIN=ER;   ER=@SUM(N(i):E(i));
 @FOR(N(i):@SUM(P1(j):IS(i,j)*VARK(j)*CHOICE(k,j))
    >= 1-BIGM*E(i));
 @FOR(P1(j):@FREE(VARK(j)));
 !  @FOR(N(i):@BIN(E(i)));
ENDSUBMODEL
```

Moreover, we define new "SUBMODEL" called "RIP2" that is one of character string of complete Revised IP-OLDF.

```
SUBMODEL RIP2:
 @FOR(N(i): @BIN(E(i)));
ENDSUBMODEL
```

Through these two "SUBMODEL," we can discriminate the data by Revised IP-OLDF with the "@SOLVE (LP, RIP2)" function instead of @SOLVE(RIP) in the "CALC" section. Next, we define Revised IPLP-OLDF. In the first stage, we discriminate the data by Revised LP-OLDF. In the second phase, Revised IP-OLDF discriminate the restricted cases misclassified by Revised LP-OLDF. Therefore, we must distinguish the two alternatives stored in the array "CONSTANT," and Revised IP-OLDF discriminates only the misclassified cases by "SUBMODEL CONS:".

```
SUBMODEL CONS:

  @FOR(N(I) | CONSTANT(i)#GT#0: @BIN(E(I)));
  @FOR(N(I) | CONSTANT(i)#EQ#0: E(I)=0);

ENDSUBMODEL
```

In the "CALC" section, we insert the statements shown below for Revised IPLP-OLDF.

```
@SOLVE(LP);
@FOR(N(i):@IFC(E(I)#EQ#0:CONSTANT(i)=0;              @ELSE
CONSTANT(i)=1;));
MNM =0; ER1=0; MNM2= 0; ER2=0;
@FOR( P1( J): VARK( J) =0; @RELEASE( VARK( J)));
@SOLVE(RIP, CONS);
```

S-SVM has two objects in Eq. (9.2). These two objects are combined by defining some "penalty c." In this research, two S-SVMs, such as SVM4 ($c = 10^4$) and SVM1 ($c = 1$) are examined in order to demonstrate that the error rate means of SVM4 are almost better than those of SVM1. If we delete the second object, "$c \times \Sigma e_i$," or set "$c = 0$," S-SVM in Eq. (9.2) becomes H-SVM.

$$\text{MIN} = ||\mathbf{b}||^2/2 + c \times \Sigma e_i \qquad (9.2)$$
$$y_i \times ({}^t\mathbf{x_i}\mathbf{b} + b_0) \geq 1 - M \times e_i$$

**b**, **x**$_i$, (${}^t$**x**$_i$ **b** + $b_0$), $y_i$: same as Eq. (9.2)
$c$: penalty c
$e_i$: non-negative decision variable

```
SUBMODEL SSVM:

MIN=ER;

ER=@SUM(P(J1): VARK(j1)^2)/2+BIGM*@SUM(N(i):E(i));

  @FOR (N(i):@SUM(P1(j):IS(i,j)*VARK(j)*CHOICE(k,j)) >= 1-E(i));

  @FOR (P1(j): @FREE(VARK(j)));
ENDSUBMODEL
```

If we insert six LDFs before the "CALC" section, we can easily discriminate the data by six LDFs.

## 9.5 Discrimination of Iris Data by LINGO

We can discriminate Iris data in Excel by Revised IP-OLDF using the "SETS, DATA, MODEL, CALC, and DATA" sections. In the "SETS" section, "$P$, $P1$, $N$, and ERR(MS)" are one-dimensional sets with element numbers of 4, 5, 100, and 15, respectively, defined in the "DATA" section. Set "$P1$" has one-dimensional array "VARK" that stores the discriminant coefficient. Set "$N$" has two one-dimensional arrays. "$E$" stores the 100 binary integer values of "$e_i$", and "CONSTANT" stores 100 discriminant scores. "ERR (MS):" has the two one-dimensional arrays. "NM" and "ZERO" store the number of misclassifications (NM) and the number of cases on the discriminant hyperplane as shown in Fig. 9.4. If we discriminate the data, the "NM" column shows MNMs of 15 models. From the "ZERO" column, we confirm Revised IP-OLDF is free from Problem 1. Because other LDFs cannot avoid Problem 1, all LDFs must output these numbers. Currently, we cannot trust the NM output of these LDFs. "VARK100" stores the 15 coefficients of Revised IP-OLDF of "VARK" shown in Fig. 9.5.

```
MODEL:
SETS:
 P; P1:VARK; P2; N: E,CONSTANT; MS;
 ERR(MS):NM,ZERO;
 D(N,P1):IS; MB(MS,P1):CHOICE;
 VP(MS,P1):VARK100;
ENDSETS
DATA:
  P=1..4; P1=1..5; N=1..100; MS=1..15;
  CHOICE,IS=@OLE();
ENDDATA
```

Here, we insert six SUBMODELs of MP-based LDFs.

```
CALC:
    @SET('DEFAULT'); @SET('TERSEO',2);
    K=1; G=1;
    LEND=@SIZE(MS);
    @WHILE(K#LE#LEND:
    @FOR( P1( J): VARK( J) = 0; @RELEASE( VARK( J)));NM=0; Z=0; BIGM=10000;
    @SOLVE(RIP); !Change the submodel name.;
    @FOR(P1(J1):VARK100(@SIZE(MS)*(G-1) +K, J1) =VARK(J1)*CHOICE(k,J1));
    @FOR(n(I):   constant(i)= @SUM(P1(J1): IS(i,J1)*VARK(J1)*CHOICE(k,J1)));
    @FOR(n(I): @IFC(CONSTANT(i) #EQ#0: Z=Z+1));
     @FOR(n(I): @IFC(CONSTANT(i) #LT#0: NM=NM+1));
    NM(K)=NM; ZERO(K)=Z; K=K+1 );
ENDCALC
DATA:
    @OLE( )=NM,ZERO,VARK100;
ENDDATA
END
```

## 9.6 How to Generate Resampling Samples and Prepare Data in Excel File

We generate resampling samples from the original data and evaluate six MP-based LDFs by Method 1. I explain this procedure with the Iris data that consist of two species, such as virginica ($y_i = 1$) and versicolor ($y_i = -1$). These species are composed of 50 cases with four variables and classifier $y_i$. We copy each species 100 times. We add a random number as the seventh variable and sort it in ascending order by "the random number ($R$ column)." We consider this data set as a pseudo-population and the validation sample that has the same statistics values, such as average or range, as the original data. Next, we divide this sample into 100 subsamples and add the subsample number (SS column) from 1 to 100 as the sixth variable. Each resampling sample consists of the 5000 cases and seven variables as listed in Table 9.1. The "$R$" column is the random number sorted by ascending order in each class. Six variables, excluding "$R$," are input by "ES = @OLE ();" in the "DATA" section." The "@OLE ()" function inputs data ES into the Excel array name, such as "A2: F10001," if cell "X1" is located in "A1," and defines the LINGO array ES. A total of 100 subsamples are the training samples, and a total resampling sample is used as the validation sample. We consider the validation sample the pseudo-population, and the training samples are the samples from the

9.6 How to Generate Resampling Samples ...

**Table 9.1** Re-sampling sample: ES

| X1 | X2 | X3 | X4 | $y_i$ | SS | R |
|---|---|---|---|---|---|---|
| x(1, 1) | x(2, 1) | x(3, 1) | x(4, 1) | 1 | 1 | |
| | | | | 1 | ... | |
| | | | | 1 | 1 | |
| | | | | 1 | ... | |
| x(1, 4951) | x(2, 4951) | x(3, 4951) | x(4, 4951) | 1 | 100 | |
| | | | | 1 | ... | |
| | | | | 1 | 100 | |
| −x(1, 5001) | −x(2, 5001) | −x(3, 5001) | −x(4, 5001) | −1 | 1 | |
| | | | | −1 | ... | |
| | | | | −1 | 1 | |
| | | | | −1 | ... | |
| | | | | −1 | 100 | |
| | | | | −1 | ... | |
| −x(1, 10000) | −x(2, 10000) | −x(3, 10000) | −x(4, 10000) | −1 | 100 | |

pseudo-population. We should set the validation sample uniquely and evaluate the 100 different training samples by the validation sample as the pseudo-population.

The two-dimensional set "MB (MS, P1)" with 15 × 5 defines the array "CHOICE" that defines the selected variable pattern of all possible models. We can control the models to discriminate by this array. The "CHOICE = @OLE()" function inputs the Excel array name "CHOICE" indicated in Table 9.2, such as "J2: N16" if the "SNI" cell is located in H1. The second row (J2: N2) shows the full model (X1, X1, X3, X4). "J3:N3" is the three-variable model (X2, X3, X4). In the "CALC" section, if we set "RIP" in the "@SOLVE (RIP);" command, we can obtain the results of Revised IP-OLDF. In the second "DATA" section, "@OLE () = IC, IC_2, EC, EC_2;" outputs MNMs and the number of a discriminant hyperplane in the training samples, and NMs and the number of the discriminant hyperplane in the validation samples. The two-dimensional set "ERR(MS, G100)" with 15 × 100 defines four arrays. The "@OLE() = IC" command outputs MNM to the Excel array called "IC." "@OLE() = VARK100, SCORE2;" outputs VARK100 with 1500 × 6 that contains 15 × 100 LDFs and SCORES with 10,000 × 15 discriminant scores in the validation samples in Excel. From IC and EC, we compute the error rate mean M1 from the training sample and M2 in the validation sample. We propose that the model with minimum value of M2 is the best model, and the model selection procedure selects the best model with minimum values of M2s among six LDFs. From "VARK100," we can compute the 95 % CI of the discriminant coefficients.

**Table 9.2** CHOICE

| SN | p | X1 | X2 | X3 | X4 | c |
|---|---|---|---|---|---|---|
| 1 | 4 | 1 | 1 | 1 | 1 | 1 |
| 2 | 3 | 0 | 1 | 1 | 1 | 1 |
| 3 | 3 | 1 | 0 | 1 | 1 | 1 |
| 4 | 3 | 1 | 1 | 0 | 1 | 1 |
| 5 | 3 | 1 | 1 | 1 | 0 | 1 |
| 6 | 2 | 0 | 0 | 1 | 1 | 1 |
| 7 | 2 | 0 | 1 | 0 | 1 | 1 |
| 8 | 2 | 1 | 0 | 0 | 1 | 1 |
| 9 | 2 | 0 | 1 | 1 | 0 | 1 |
| 10 | 2 | 1 | 0 | 1 | 0 | 1 |
| 11 | 2 | 1 | 1 | 0 | 0 | 1 |
| 12 | 1 | 0 | 0 | 0 | 1 | 1 |
| 13 | 1 | 0 | 0 | 1 | 0 | 1 |
| 14 | 1 | 0 | 1 | 0 | 0 | 1 |
| 15 | 1 | 1 | 0 | 0 | 0 | 1 |

## 9.7 Set Model by LINGO

Fisher never formulated the equation for SE of error rates and discriminant coefficients. If we discriminate the data by Method 1, we can calculate the 95 % CI of the error rate and discriminant coefficient. We obtain the Philosopher's Stone. The "SET" section defines eight one-dimensional sets, such as P, P1, P2, N, N2, MS, MS100, and G100. "P, P1, and P2" are the number of independent variables, number of (independent variables + intercept), and the number of (independent variables + intercept + subsample), respectively. These figures of the elements in the "DATA" section are 4, 5, and 6, respectively. Only "P1" defines the one-dimensional array called "VARK" with five elements that store the discriminant coefficients of the training sample. Set "N" with 100 elements defines "E, SCORE, CONSTANT." Array "E" is a one-dimensional array with 100 elements and corresponds to "$e_i$." "N:; N2:; MS:; MS100:; G100:;" are one-dimensional sets with 100 elements, 10,000 elements, 15 elements, 1500 elements, and 100 elements, respectively. The two-dimensional set "D(N, P1):" with 100 × 5 has the same size array "IS" that stores the 100 subsamples as the evaluation samples. "D2 (N2, P2):" with 10,000 × 6 has the same size array "ES" that stores the resampling-sample as the validation sample. In the "CALC" section, if we set "@SOLVE (RIP);", we can discriminate the Iris data and output six results to the Excel arrays. "IC and EC" are the 100 MNMs in the training sample and 100 NMs in the validation sample. "IC_2 and EC_2" are the 100 numbers on the discriminant hyperplane in the training and the validation samples. From these figures, we calculate the error rate means, such as "M1 and M2", from the training and validation samples. "VARK100" is the 1500 discriminant coefficients of 15 models. We can calculate the 95 % CI of the discriminant coefficients. "SCORE2" is the

## 9.7 Set Model by LINGO

10,000 discriminant scores. We will analyze this information in near future because the priority is not high.

```
MODEL: The Method 1 for iris data;
SETS:
  P; P2; P1: VARK;
  N : E, SCORE, CONSTANT; N2: SCORE2; MS : ; MS100 : ;  G100 :;
  D (N, P1): IS;
  D2 (N2, P2): ES;
  MB (MS, P1): CHOICE;
  ERR(MS,G100): IC, IC_2, EC, EC_2;
  SS(N2,MS): SCORE2;
  VVV(MS100, P2): VARK100;
ENDSETS
DATA:
    P=1..4; P1=1..5; P2=1..6;
    N=1..100; N2=1..10000;
    MS=1..15; G100=1..100; MS100=1..1500 ;
    BIGM=10000;
    CHOICE=@OLE(); ES=@OLE();
ENDDATA
```

Here, we insert the six LDFs described in Chap. 2.

```
CALC:
!Reset all options to default; @SET('DEFAULT');
!@SET('TERSEO',1);!Allow for minimal output;
@SET('TERSEO',2);
!Global solver (1:yes, 0:no); @SET('GLOBAL',1);
!Quadratic recognition (1:yes, 0:no);@SET('USEQPR',1);
!Multisarts (1:Off, >1 number of starts); @SET
('MULTIS',1);
!Number of threads; !@SET('THRDS',4);
!Print output immediately (1:yes, 0:no); @SET
('OROUTE',1);
!No need to compute dual values; @SET('DUALCO',0);
```

## 9 LINGO Program 1 of Method 1

```
  K=1; Lend=@SIZE(MS);
@WHILE (K#LE#Lend: f=1;
@WHILE (f#LE#100:
    @FOR(D(i, j): IS(i, j)=ES( @SIZE(N)*(f-1)+i, j));
  MNM=0; ER1=0;MNM2=0;ER2=0;
@FOR( P1( J): VARK( J) = 0;@RELEASE( VARK( J)));

    @SOLVE(RIP);! Set the submodel name here;

    @FOR(P1(j):VARK100(100*(k-1)+f,j)=VARK(j));
     VARK100 (100*(k-1)+f, @SIZE(P2))=K;
@FOR(n(1):SCORE(1)=@SUM(P1(j):IS(1,j)*VARK(j)*
   CHOICE (k, j)));
@FOR(n2(nn):SCORE2(nn,K)=@SUM(P1(j):ES(NN, j)* VARK(j)*CHOICE(k, j)));
 @FOR(n(1):@IFC(SCORE(1)#LT#0:  MNM=MNM+1));
@FOR(n2(nn):@IFC(SCORE2(nn, k)#LT#0:ER1=ER1+1 ));
 @FOR(n(1):@IFC(SCORE(1)#EQ#0: MNM2=MNM2+1));
@FOR(n2(nn):@IFC(SCORE2(nn,k)#EQ#0:ER2=ER2+1 );
  IC(K,f)=MNM;EC(k,f)=ER1;
  IC_2(K,f)=MNM2;
  EC_2(K,f)=ER2;
  f=f+1);
ENDCALC
```

```
DATA:
   @OLE() = IC, EC, IC_2, EC_2;
   @OLE() = VARK100, SCORE2;
ENDDATA
END
```

# Index

**A**
Akaike's Information Criterion (AIC), 5, 20, 38, 41, 61, 69, 145, 158, 177
All possible combination models, 5, 9, 50, 99, 178

**B**
Basic gene set(s) (BGS(s)), 31, 139, 141, 154, 166, 168, 173, 174, 177–179, 186
Basic gene set or subspace, 2
Bayesian Information Criteria (BIC), 5, 25, 38, 41, 69, 118, 158, 177
Best model, 2, 9, 25, 27, 30, 38, 48, 50, 53, 57, 66, 78, 95, 97, 99, 107, 111, 112, 117, 118, 126, 128, 150, 186, 201
Box–whisker plots, 139
By the mean of error rates of the validation sample, 96, 122, 163, 191

**C**
Cancer diseases, 2
Catch-up modeling, 12
Caused problem 3, 82
Cephalo-Pelvic Disproportion (CPD) data, 1, 15, 30, 50, 57, 58, 64, 69
Coefficient standard, 117, 118, 128
Collinearities, 1, 30, 57, 58, 60, 69, 72, 78
Convex Polyhedron(s) (CP(s)), 14, 15, 17, 18, 28, 81, 85

**D**
Data Envelopment Analysis (DEA), 139
Dataset, The, 2, 5, 10, 15, 28, 81, 82, 140, 143, 164–166, 177, 178, 183, 184
Defect of IP-OLDF, 1, 15, 30
Determine pass/fail using examination scores, 3, 4, 8, 9, 20, 30, 99, 100, 113, 117
Diff, 26, 52, 103, 107
Diff1, 42, 69, 74, 86, 93, 102, 106, 110, 143

Discriminant coefficients, 5, 8, 9, 12, 25, 52, 100, 104, 117, 145, 202
Discriminant rule, 3, 22, 100, 150

**E**
Earth model, 7
18 pass/fail determinations using examination scores, 1, 5, 30
Error rate means, 2, 50, 99, 103, 118, 147, 198, 202
Excel add-in solver, 84
Explanation 1 of Matroska feature-selection method, 167

**F**
Feature (model or variable) selection, 14
Feature-selection of the dataset(s), 6, 28
Features naturally, 2, 6, 31, 82, 97, 150, 166
Fisher, R., 1, 10
Fisher's assumption, 6, 7, 10, 21, 23, 26, 29, 38, 41, 57, 78
Fisher's LDF, 1, 3–7, 9–11, 20, 24, 26, 38, 39, 43, 66, 71, 85, 96, 100, 103, 110, 112, 147, 191
Five serious problems, 1, 3, 163
100-fold cross-validation for small sample method, 2, 37, 48, 57, 82, 122
For the training and validation samples, 2, 9, 25, 38, 99, 103, 118, 147, 191, 202
Forward and backward stepwise procedure(s), 15, 41, 57

**G**
Gene analysis, 2, 4, 14, 28, 31, 82, 118, 119, 130, 143, 167, 179, 186
Generalized inverse-matrices technique, 163
Global minimum solution, 87, 192
Glover, L., 11

© Springer Science+Business Media Singapore 2016
S. Shinmura, *New Theory of Discriminant Analysis After R. Fisher*,
DOI 10.1007/978-981-10-2164-0

Good generalization ability, 12, 51, 52, 103, 107

**H**
Hard-margin SVM (H-SVM), 1, 12, 28, 37, 103, 113, 120, 128, 132, 150, 163, 166, 196
Heuristic OLDF, 57
High correlation, 58

**I**
Inferential statistics, 9–11, 20, 21, 25, 41, 53
IP-OLDF, 100, 107, 110, 113, 117, 122, 126, 128, 130, 134, 148, 150, 163, 168, 186, 199
Iris data, 7, 29, 37, 38, 40, 44, 78

**J**
Japanese-automobile data, 2, 4, 30, 62, 118, 139, 150, 164
JMP, 4, 5, 38, 63, 99, 129, 164

**K**
K-fold cross-validation for small sample method, 99, 117

**L**
LASSO, 2, 28, 82, 118, 140, 165, 168, 176
Leave-one-out (LOO), 2
Leave-one-out procedure (LOO procedure), 57, 117, 163
LINDO Systems Inc, 25, 37, 191, 196
Linear discriminant function (LDF), 99
Linearly separable data (LSD), 1, 21, 40, 81, 82, 97, 100
Linearly separable dataset (LSD), 163
LINGO, 9, 28, 37, 38, 45, 87, 99, 121, 164, 167, 177, 192, 200
LINGO k-best option, 62, 85
LINGO model, 29, 37, 45
LINGO model for six MP-based LDFs, including H-SVM, 9, 38, 44, 50, 147, 164
Liitschwager, J.M., 12
Logistic regression, 1, 7, 9, 23, 38, 42, 52, 64, 71, 89, 101, 102, 106, 113, 122, 135, 164, 167, 177
LOO procedure, 9, 25, 99, 145, 191

**M**
M1(s) and M2(s), 71, 93, 103, 147
M1, 2, 9, 26, 38, 51, 99, 107, 201
M2, 2, 9, 38, 99, 135, 163, 191, 201
M1 & M2 are the mean error rates in the training and validation samples, 51, 97, 99, 122, 163, 191
Mahalanobis distance, 7, 8, 10
Mahalanobis–Taguchi method, 10
Many OCPs, 30, 58, 62, 66, 69, 78
Markowitz portfolio model, 13
Matroska, 2, 29, 118, 119, 130, 132, 140, 154, 167
Matroska feature-selection method, 82
Matroska feature-selection method for microarray dataset, 2, 31, 118, 163
Matroska product, 130, 132, 152
Matroska series, 132, 168, 174, 178
Matroska structure, 2, 15, 118, 119, 130, 140, 183
Maximum t-value standard, 117, 118, 128
Medical diagnosis, 3, 4, 7, 18, 100, 179
Method 1, 2, 9, 10, 25, 38, 66, 82, 97, 112, 124, 147, 186, 191
Method 2, 2, 6, 28, 31, 118, 132, 140, 152, 155, 164, 169
Minimum error rate mean in the validation samples, 117
Minimum M2 standard, 117, 126, 127, 135, 150
Minimum number of misclassifications (minimum NM, MNM), 1, 58, 81, 117, 139, 163
MNM criterion, 2, 8, 12, 20, 97, 166
MNM decreases monotonously, 14, 17, 28, 64, 69, 74, 140
$MNM_k = 0$, 14, 17, 28
$MNM_k >= MNM_{(k+1)}$, 118
$MNM_q >= MNM_{(q+1)}$, 119
Model selection procedure, 2, 5, 9, 25, 29, 38, 41, 53, 57, 78, 145, 163, 201
Monotonic decrease of MNM, 8, 9, 42, 132, 141

**N**
New model selection procedure such as the best model, 5
New theory of discriminant analysis, 1, 135, 163
95 % CI, 5, 9, 25, 30, 52, 66, 68, 99, 129, 151
95 % CI of discriminant coefficients, 25, 30, 38, 50, 58, 69, 95, 100, 117, 191, 202
95 % CI of error rates, 9, 117

# Index

95 % CI of the error rate and discriminant coefficient, 29, 48, 50, 163, 186, 202
Number of Misclassifications (NM), 3, 199

## O

Optimal CP (OCP), 14, 17, 62, 81, 85, 128
Optimal linear discriminant function using integer programming *See*, IP-OLDF, 81, 117
Ordinary statistical methods, 6, 28, 165, 166, 173, 178, 183
Overlapping data, 12, 120, 179

## P

Pass/fail determination, 4, 9, 23, 62, 112, 148
Pass/fail determination of examination scores, 21, 29
Pattern recognition, 7, 13, 179
Penalty c = 1, 1, 37, 120, 167
Penalty c = 10,000, 1, 37, 167
Plug-in rule1, 5, 20, 41, 91, 145, 151
Plug-in rule2, 10
Portfolio selection, 12
Principal Component Analysis (PCA), 6, 69, 82, 144, 166, 183
Problem 1, 1–3, 8, 14–18, 22, 42, 81, 84, 86, 89, 93, 100, 102, 106, 107, 143, 151, 163, 199
Problem 2, 2, 4, 12, 20, 71, 81, 102, 106, 117, 120, 163
Problem 3, 2, 4, 5, 8, 11, 22, 24, 81, 82, 139, 141, 163
Problem 4, 2, 5, 8, 10, 25, 38, 48, 81, 82, 91, 99, 117, 163
Problem 5, 2, 5, 9, 24, 28, 81, 82, 164

## Q

Quadratic discriminant function (QDF), 1, 57, 82, 139
Quadratic programming (QP), 12
Quality control, 7, 10

## R

Regularized discriminant analysis (RDA), 4, 82, 120, 140
Revised IPLP-OLDF, 1, 3, 9, 16, 18, 26, 37, 42, 52, 53, 57, 64, 66, 74, 82, 84, 86, 89, 100, 102, 106, 110, 117, 120, 121, 124, 129, 134, 143, 147, 151, 164, 166, 167, 196–198

Revised IP-OLDF, 1–3, 5, 8, 9, 12, 15–18, 20, 21, 23, 24, 26–28, 37, 38, 42, 46, 52, 57, 58, 65, 68, 71, 72, 81, 82, 84, 85, 89, 91, 95, 97, 100, 101, 103, 107, 110, 111, 117, 120, 122, 124, 128, 130, 132, 139, 140, 143, 147, 148, 150, 152, 154, 163–167, 174, 177, 191, 194, 196, 197, 199, 201
Revised IP-OLDF based on MNM, 3, 24, 72, 120, 143
Rubin, P.A, 11

## S

Scatter plot, 18, 39, 81, 101, 119, 120, 128, 181, 185
SET and CALC notation, 192
SETS section, 44, 193, 199
Six microarrays datasets, 2, 5, 28
Small Matroska(s) (SM(s)), 2, 132, 134, 140, 164
S-SVM, 13, 26, 120, 132, 165, 198
Stam, A., 11, 165
Standard deviation (SD), 48
Standard errors (SEs), 2, 129
Statistical discriminant theory, 6
Stepwise procedure, 5
Structure of a Matroska, 164
Student data, 1, 3, 8, 15, 18, 62, 81, 82, 93
Support vector (SV), 12, 128, 167
Support Vector Machine (SVM), 46, 57, 100, 117
SVM1, 1, 3, 9, 13, 26, 37, 43, 53, 58, 66, 74, 86, 89, 93, 96, 100, 102, 104, 106, 110, 120, 124, 129, 132, 143, 147, 151, 167, 196, 198
SVM4, 1, 3, 9, 13, 18, 26, 27, 37, 43, 46, 53, 64, 66, 74, 86, 88, 89, 93, 96, 100, 102, 106, 110, 120, 122, 129, 132, 134, 143, 147, 151, 155, 167, 196, 198
Swiss banknote, 5, 62, 118, 165
Swiss banknote data, 1, 2, 4, 5, 9, 12, 14, 15, 17, 20, 21, 24, 29, 53, 81, 82, 117–119, 122, 129, 130, 132, 164, 180

## T

Theory, the, 1, 3, 9, 16, 28, 38, 57, 69, 81, 82, 99, 152, 164, 168
Traditional inferential statistics, 9, 10, 99
Training and validation samples, 9, 48, 99
Trivial LDF(s), 3, 9, 21, 22, 29, 30, 100, 108, 111, 117
Trivial LSD, 4

True OCPs, 84
Two error rate, 2, 118
Two means, 6, 9, 38, 99, 118, 163
Two S-SVMs, 13, 164, 196, 198

**U**
Unresolved problem, 3, 17
Useful model, 2, 5, 8, 11, 18, 44, 50, 52, 53, 57, 69, 81, 82, 97, 100, 105, 117, 127, 151, 165, 167, 169, 186

**V**
Variance–covariance matrices, 6, 7, 10, 23, 49, 120, 143, 163
Various ratings, 7

**W**
Wang, C, 12
What's Best!, 18, 191

CPSIA information can be obtained
at www.ICGtesting.com
Printed in the USA
LVOW02*1023080117
520185LV00001B/54/P

9 789811 021633